SOURCE

The Prentice Hall
ENGINEERING SOURCE

Introduction to
C++

Delores M. Etter

Department of Electrical and Computer Engineering
University of Colorado, Boulder

Prentice Hall
Upper Saddle River, NJ 07458

Library of Congress Cataloging-In-Publication Data

Etter, Delores Maria.
    Introduction to C++/ Delores M. Etter
       p.   cm.
    Includes bibliographical references and index.
    ISBN 0–13–011855-9
    1. C (Computer program language)   I. Title.
QA76.73.C15E853   1998
005.13 '3—dc21

                                      98-50974
                                          CIP

Editor-in-chief: **MARCIA HORTON**
Acquisitions editor: **ERIC SVENDSEN**
Director of production and manufacturing: **DAVID W. RICCARDI**
Managing editor: **EILEEN CLARK**
Editorial/production supervision: **ROSE KERNAN**
Cover director: **JAYNE CONTE**
Creative director: **AMY ROSEN**
Marketing manager: **DANNY HOYT**
Manufacturing buyer: **PAT BROWN**
Editorial assistant: **GRIFFIN CABLE**

The author and publisher of this book have used their best efforts in
preparing this book. These efforts include the development, research,
and testing of the theories and programs to determine their effective-
ness. The author and publisher shall not be liable in any event for inci-
dental or consequential damages in connection with, or arising out of,
the furnishing, performance, or use of these programs.
Printed in the United States of America

MATLAB is a registered trademark of the MathWorks, Inc.

10 9 8 7 6 5 4 3 2 1

ISBN 0-13-011855-9

Prentice-Hall International (UK) Limited, *London*
Prentice-Hall of Australia Pty. Limited, *Sydney*
Prentice-Hall Canada, Inc., *Toronto*
Prentice-Hall Hispanoamericana, S.A., *Mexico*
Prentice-Hall of India Private Limited, *New Delhi*
Prentice-Hall of Japan, Inc., *Tokyo*
Simon & Schuster Asia Pte., Ltd., *Singapore*
Editora Prentice-Hall do Brazil, Ltda., *Rio de Janeiro*

Contents

6 CHARACTER DATA

About ESource

The Challenge

Professors who teach the Introductory/First-Year Engineering course popular at most engineering schools have a unique challenge—teaching a course defined by a changing curriculum. The first-year engineering course is different from any other engineering course in that there is no real cannon that defines the course content. It is not like Engineering Mechanics or Circuit Theory where a consistent set of topics define the course. Instead, the introductory engineering course is most often defined by the creativity of professors and students, and the specific needs of a college or university each semester. Faculty involved in this course typically put extra effort into it, and it shows in the uniqueness of each course at each school.

Choosing a textbook can be a challenge for unique courses. Most freshmen require some sort of reference material to help them through their first semesters as a college student. But because faculty put such a strong mark on their course, they often have a difficult time finding the right mix of materials for their course and often have to go without a text, or with one that does not really fit. Conventional textbooks are far too static for the typical specialization of the first-year course. How do you find the perfect text for your course that will support your students educational needs, but give you the flexibility to maximize the potential of your course?

ESource—The Prentice Hall Engineering Source
http://emissary.prenhall.com/esource

Prentice Hall created ESource—The Prentice-Hall Engineering Source—to give professors the power to harness the full potential of their text and their freshman/first year engineering course. In today's technologically advanced world, why settle for a book that isn't perfect for your course? Why not have a book that has the exact blend of topics that you want to cover with your students?

More then just a collection of books, ESource is a unique publishing system revolving around the ESource website—http://emissary.prenhall.com/esource/. ESource enables you to put your stamp on your book just as you do your course. It lets you:

Control   You choose exactly what chapters or sections are in your book and in what order they appear. Of course, you can choose the entire book if you'd like and stay with the authors original order.

Optimize   Get the most from your book and your course. ESource lets you produce the optimal text for your students needs.

Customize   You can add your own material anywhere in your text's presentation, and your final product will arrive at your bookstore as a professionally formatted text.

ESource Content

All the content in ESource was written by educators specifically for freshman/first-year students. Authors tried to strike a balanced level of presentation, one that was not either too formulaic and trivial, but not focusing heavily on advanced topics that most introductory students will not encounter until later classes. A developmental editor reviewed the books and made sure that every text was written at the appropriate level, and that the books featured a balanced presentation. Because many professors do not have extensive time to cover these topics in the classroom, authors prepared each text with the idea that many students would use it for self-instruction and independent study. Students should be able to use this content to learn the software tool or subject on their own.

While authors had the freedom to write texts in a style appropriate to their particular subject, all followed certain guidelines created to promote the consistency a text needs. Namely, every chapter opens with a clear set of objectives to lead students into the chapter. Each chapter also contains practice problems that tests a student's skill at performing the tasks they have just learned. Chapters close with extra practice questions and a list of key terms for reference. Authors tried to focus on motivating applications that demonstrate how engineers work in the real world, and included these applications throughout the text in various chapter openers, examples, and problem material. Specific Engineering and Science **Application Boxes** are also located throughout the texts, and focus on a specific application and demonstrating its solution.

Because students often have an adjustment from high school to college, each book contains several **Professional Success Boxes** specifically designed to provide advice on college study skills. Each author has worked to provide students with tips and techniques that help a student better understand the material, and avoid common pitfalls or problems first-year students often have. In addition, this series contains an entire book titled *Engineering Success* by Peter Schiavone of the University of Alberta intended to expose students quickly to what it takes to be an engineering student.

Creating Your Book

Using ESource is simple. You preview the content either on-line or through examination copies of the books you can request on-line, from your PH sales rep, or by calling(1-800-526-0485). Create an on-line outline of the content you want in the order you want using ESource's simple interface. Either type or cut and paste your own material and insert it into the text flow. You can preview the overall organization of the text you've created at anytime (please note, since this preview is immediate, it comes unformatted.), then press another button and receive an order number for your own custom book . If you are not ready to order, do nothing—ESource will save your work. You can come back at any time and change, re-arrange, or add more material to your creation. You are in control. Once you're finished and you have an ISBN, give it to your bookstore and your book will arrive on their shelves six weeks after the order. Your custom desk copies with their instructor supplements will arrive at your address at the same time.

To learn more about this new system for creating the perfect textbook, go to **http://emissary.prenhall.com/esource/**. You can either go through the on-line walkthrough of how to create a book, or experiment yourself.

Community

ESource has two other areas designed to promote the exchange of information among the introductory engineering community, the Faculty and the Student Centers. Created and maintained with the help of Dale Calkins, an Associate Professor at the University of Washington, these areas contain a wealth of useful information and tools. You can preview outlines created by other schools and can see how others organize their courses. Read a monthly article discussing important topics in the curriculum. You can post your own material and share it with others, as well as use what others have posted in your own documents. Communicate with our authors about their books and make suggestions for improvement. Comment about your course and ask for information from others professors. Create an on-line syllabus using our custom syllabus builder. Browse Prentice Hall's catalog and order titles from your sales rep. Tell us new features that we need to add to the site to make it more useful.

Supplements

Adopters of ESource receive an instructor's CD that includes solutions as well as professor and student code for all the books in the series. This CD also contains approximately **350 Powerpoint Transparencies** created by Jack Leifer of the University South Carolina—Aiken. Professors can either follow these transparencies as pre-prepared lectures or use them as the basis for their own custom presentations. In addition, look to the web site to find materials from other schools that you can download and use in your own course.

Titles in the ESource Series

Introduction to Unix
0-13-095135-8
David I. Schwartz

Introduction to Maple
0-13-095133-1
David I. Schwartz

Introduction to Word
0-13-254764-3
David C. Kuncicky

Introduction to Excel
0-13-254749-X
David C. Kuncicky

Introduction to MathCAD
0-13-937493-0
Ronald W. Larsen

Introduction to AutoCAD, R. 14
0-13-011001-9
Mark Dix and Paul Riley

Introduction to the Internet, 2/e
0-13-011037-X
Scott D. James

Design Concepts for Engineers
0-13-081369-9
Mark N. Horenstein

Engineering Design—A Day in the Life
of Four Engineers
0-13-660242–8
Mark N. Horenstein

Engineering Ethics
0-13-784224-4
Charles B. Fleddermann

Engineering Success
0-13-080859-8
Peter Schiavone

Mathematics Review
0-13-011501-0
Peter Schiavone

Introduction to C
0-13-011854-0
Delores M. Etter

Introduction to C++
0-13-011855-9
Delores M. Etter

Introduction to MATLAB
0-13-013149-0
Delores M. Etter with David C. Kuncicky

Introduction to FORTRAN 90
0-13-013146-6
Larry Nyhoff and Sanford Leestma

About the Authors

N o project could ever come to pass without a group of authors who have the vision and the courage to turn a stack of blank paper into a book. The authors in this series worked diligently to produce their books, provide the building blocks of the series.

Delores M. Etter is a Professor of Electrical and Computer Engineering at the University of Colorado. Dr. Etter was a faculty member at the University of New Mexico and also a Visiting Professor at Stanford University. Dr. Etter was responsible for the Freshman Engineering Program at the University of New Mexico and is active in the Integrated Teaching Laboratory at the University of Colorado. She was elected a Fellow of the Institute of Electrical and Electronic Engineers for her contributions to education and for her technical leadership in digital signal processing. IN addition to writing best-selling textbooks for engineering computing, Dr. Etter has also published research in the area of adaptive signal processing.

Sanford Leestma is a Professor of Mathematics and Computer Science at Calvin College, and received his Ph.D from New Mexico State University. He has been the long time co-author of successful textbooks on Fortran, Pascal, and data structures in Pascal. His current research interests are in the areas of algorithms and numerical computation.

Larry Nyhoff is a Professor of Mathematics and Computer Science at Calvin College. After doing bachelors work at Calvin, and Masters work at Michigan, he received a Ph.D. from Michigan State and also did graduate work in computer science at Western Michigan. Dr. Nyhoff has taught at Calvin for the past 34 years—mathematics at first and computer science for the past several years. He has co-authored several computer science textbooks since 1981 including titles on Fortran and C++, as well as a brand new title on Data Structures in C++.

Acknowledgments: We express our sincere appreciation to all who helped in the preparation of this module, especially our acquisitions editor Alan Apt, managing editor Laura Steele, development editor Sandra Chavez, and production editor Judy Winthrop. We also thank Larry Genalo for several examples and exercises and Erin Fulp for the Internet address application in Chapter 10. We appreciate the insightful review provided by Bart Childs. We thank our families—Shar, Jeff, Dawn, Rebecca, Megan, Sara, Greg, Julie, Joshua, Derek, Tom, Joan; Marge, Michelle, Sandy, Lori, Michael—for being patient and understanding. We thank God for allowing us to write this text.

Mark Dix began working with AutoCAD in 1985 as a programmer for CAD Support Associates, Inc. He helped design a system for creating estimates and bills of material directly from AutoCAD drawing databases for use in the automated conveyor industry. This system became the basis for systems still widely in use today. In 1986 he began collaborating with Paul Riley to create AutoCAD training materials, combining Riley's background in industrial design and training with Dix' s background in writing, curriculum development, and programming. Dix and Riley have created tutorial and teaching methods for every AutoCAD release since Version 2.5. Mr. Dix has a Master of Arts in Teaching from Cornell University and a Masters of Education from the University of Massachusetts. He is currently the Director of Dearborn Academy High School in Arlington, Massachusetts.

Paul Riley is an author, instructor, and designer specializing in graphics and design for multimedia. He is a founding partner of CAD Support Associates, a contract service and professional training organization for computer-aided design. His 15 years of business experience and 20 years of teaching experience are supported by degrees in education and computer science. Paul has taught AutoCAD at the University of Massachusetts at Lowell and is presently teaching AutoCAD at Mt. Ida College in Newton, Massachusetts. He has developed a program, <u>Computer-Aided Design for Professionals</u> that is highly regarded by corporate clients and has been an ongoing success since 1982.

David I. Schwartz is a Lecturer at SUNY-Buffalo who teaches freshman and first-year engineering, and has a Ph.D from SUNY-Buffalo in Civil Engineering. Schwartz originally became interested in Civil engineering out of an interest in building grand structures, but has also pursued other academic interests including artificial intelligence and applied mathematics. He became interested in Unix and Maple through their application to his research, and eventually jumped at the chance to teach these subjects to students. He tries to teach his students to become incremental learners and encourages frequent practice to master a subject, and gain the maturity and confidence to tackle other subjects independently. In his spare time, Schwartz is an avid musician and plays drums in a variety of bands.

Acknowledgments: I would like to thank the entire School of Engineering and Applied Science at the State University of New York at Buffalo for the opportunity to teach not only my students, but myself as well; all my EAS140 students, without whom this book would not be possible—thanks for slugging through my lab packets; Andrea Au, Eric Svendsen, and Elizabeth Wood at Prentice Hall for advising and encouraging me as well as wading through my blizzard of e-mail; Linda and Tony for starting the whole thing in the first place; Rogil Camama, Linda Chattin, Stuart Chen, Jeffrey Chottiner, Roger Christian, Anthony Dalessio, Eugene DeMaitre, Dawn Halvorsen, Thomas Hill, Michael Lamanna, Nate "X" Patwardhan, Durvejai Sheobaran, "Able" Alan Somlo, Ben Stein, Craig Sutton, Barbara Umiker, and Chester "JC" Zeshonski for making this book a reality; Ewa Arrasjid, "Corky" Brunskill, Bob Meyer, and Dave Yearke at "the Department Formerly Known as ECS" for all their friendship, advice, and respect; Jeff, Tony, Forrest, and Mike for the interviews; and, Michael Ryan and Warren Thomas for believing in me.

Ronald W. Larsen is an Associate Professor in Chemical Engineering at Montana State University, and received his Ph.D from the Pennsylvania State University. Larsen was initially attracted to engineering because he felt it was a serving profession, and because engineers are often called on to eliminate dull and routine tasks. He also enjoys the fact that engineering rewards creativity and presents constant challenges. Larsen feels that teaching large sections of students is one of the most challenging tasks he has ever encountered because it enhances the importance of effective communication. He has drawn on a two year experince teaching courses in Mongolia through an interpreter to improve his skills in the classroom. Larsen sees software as one of the changes that has the potential to radically alter the way engineers work, and his book Introduction to Mathcad was written to help young engineers prepare to be productive in an ever-changing workplace.

Acknowledgments: To my students at Montana State University who have endured the rough drafts and typos, and who still allow me to experiment with their classes—my sincere thanks.

Peter Schiavone is a professor and student advisor in the Department of Mechanical Engineering at the University of Alberta. He received his Ph.D. from the University of Strathclyde, U.K. in 1988. He has authored several books in the area of study skills and academic success as well as numerous papers in scientific research journals.

Before starting his career in academia, Dr. Schiavone worked in the private sector for Smith's Industries (Aerospace and Defence Systems Company) and Marconi Instruments in several different areas of engineering including aerospace, systems and software engineering. During that time he developed an interest in engineering research and the applications of mathematics and the physical sciences to solving real-world engineering problems.

His love for teaching brought him to the academic world. He founded the first Mathematics Resource Center at the University of Alberta: a unit designed specifically to teach high school students the necessary survival skills in mathematics and the physical sciences required for first-year engineering. This led to the Students' Union Gold Key award for outstanding contributions to the University and to the community at large.

Dr. Schiavone lectures regularly to freshman engineering students, high school teachers, and new professors on all aspects of engineering success, in particular, maximizing students' academic performance. He wrote the book *Engineering Success* in order to share with you the *secrets of success in engineering study*: the most effective, tried and tested methods used by the most successful engineering students.

Acknowledgments: I'd like to acknowledge the contributions of: Eric Svendsen, for his encouragement and support; Richard Felder for being such an inspiration; the many students who shared their experiences of first-year engineering—both good and bad; and finally, my wife Linda for her continued support and for giving me Conan.

Scott D. James is a staff lecturer at Kettering University (formerly GMI Engineering & Management Institute) in Flint, Michigan. He is currently pursuing a Ph.D. in Systems Engineering with an emphasis on software engineering and computer-integrated manufacturing. Scott decided on writing textbooks after he found a void in the books that were available. "I really wanted a book that showed how to do things in good detail but in a clear and concise way. Many of the books on the market are full of fluff and force you to dig out the really important facts." Scott decided on teaching as a profession after several years in the computer industry. "I thought that it was really important to know what it was like outside of academia. I wanted to provide students with classes that were up to date and provide the information that is really used and needed."

Acknowledgments: Scott would like to acknowledge his family for the time to work on the text and his students and peers at Kettering who offered helpful critique of the materials that eventually became the book.

David C. Kuncicky is a native Floridian. He earned his Baccalaureate in psychology, Master's in computer science, and Ph.D. in computer science from Florida State University. Dr. Kuncicky is the Director of Computing and Multimedia Services for the FAMU-FSU College of Engineering. He also serves as a faculty member in the Department of Electrical Engineering. He has taught computer science and computer engineering courses for the past 15 years. He has published research in the areas of intelligent hybrid systems and neural networks. He is actively involved in the education of computer and network system administrators and is a leader in the area of technology-based curriculum delivery.

Acknowledgments: Thanks to Steffie and Helen for putting up with my late nights and long weekends at the computer. Thanks also to the helpful and insightful technical reviews by the following people: Jerry Ralya, Kathy Kitto of Western Washington University, Avi Singhal of Arizona State University, and Thomas Hill of the State University of New York at Buffalo. I appreciate the patience of Eric Svendsen and Rose Kernan of Prentice Hall for gently guiding me through this project. Finally, thanks to Dean C.J. Chen for providing continued tutelage and support.

Mark Horenstein is an Associate Professor in the Electrical and Computer Engineering Department at Boston University. He received his Bachelors in Electrical Engineering in 1973 from Massachusetts Institute of Technology, his Masters in Electrical Engineering in 1975

from University of California at Berkeley, and his Ph.D. in Electrical Engineering in 1978 from Massachusetts Institute of Technology. Professor Horenstein's research interests are in applied electrostatics and electromagnetics as well as microelectronics, including sensors, instrumentation, and measurement. His research deals with the simulation, test, and measurement of electromagnetic fields. Some topics include electrostatics in manufacturing processes, electrostatic instrumentation, EOS/ESD control, and electromagnetic wave propagation.

Professor Horenstein designed and developed a class at Boston University, which he now teaches entitled Senior Design Project (ENG SC 466). In this course, the student gets real engineering design experience by working for a virtual company, created by Professor Horenstein, that does real projects for outside companies—almost like an apprenticeship. Once in "the company" (Xebec Technologies), the student is assigned to an engineering team of 3-4 persons. A series of potential customers are recruited, from which the team must accept an engineering project. The team must develop a working prototype deliverable engineering system that serves the need of the customer. More than one team may be assigned to the same project, in which case there is competition for the customer's business.

Acknowledgements: Several individuals contributed to the ideas and concepts presented in Design Principles for Engineers. The concept of the Peak Performance design competition, which forms a cornerstone of the book, originated with Professor James Bethune of Boston University. Professor Bethune has been instrumental in conceiving of and running Peak Performance each year and has been the inspiration behind many of the design concepts associated with it. He also provided helpful information on dimensions and tolerance. Several of the ideas presented in the book, particularly the topics on brainstorming and teamwork, were gleaned from a workshop on engineering design help bi-annually by Professor Charles Lovas of Southern Methodist University. The principles of estimation were derived in part from a freshman engineering problem posed by Professor Thomas Kincaid of Boston University.

I would like to thank my family, Roxanne, Rachel, and Arielle, for giving me the time and space to think about and write this book. I also appreciate Roxanne's inspiration and help in identifying examples of human/machine interfaces.

Dedicated to Roxanne, Rachel, and Arielle

Charles B. Fleddermann is a professor in the Department of Electrical and Computer Engineering at the University of New Mexico in Albuquerque, New Mexico. He is a third generation engineer—his grandfather was a civil engineer and father an aeronautical engineer—so "engineering was in my genetic makeup." The genesis of a book on engineering ethics was in the ABET requirement to incorporate ethics topics into the undergraduate engineering curriculum. "Our department decided to have a one-hour seminar course on engineering ethics, but there was no book suitable for such a course." Other texts were tried the first few times the course was offered, but none of them presented ethical theory, analysis, and problem solving in a readily accessible way. "I wanted to have a text which would be concise, yet would give the student the tools required to solve the ethical problems that they might encounter in their professional lives."

Reviewers

ESource benefited from a wealth of reviewers who on the series from its initial idea stage to its completion. Reviewers read manuscripts and contributed insightful comments that helped the authors write great books. We would like to thank everyone who helped us with this project.

Concept Document
Naeem Abdurrahman- University of Texas, Austin
Grant Baker- University of Alaska, Anchorage
Betty Barr- University of Houston
William Beckwith- Clemson University
Ramzi Bualuan- University of Notre Dame
Dale Calkins- University of Washington
Arthur Clausing- University of Illinois at Urbana-Champaign
John Glover- University of Houston
A.S. Hodel- Auburn University
Denise Jackson- University of Tennessee, Knoxville
Kathleen Kitto- Western Washington University
Terry Kohutek- Texas A&M University
Larry Richards- University of Virginia
Avi Singhal- Arizona State University
Joseph Wujek- University of California, Berkeley
Mandochehr Zoghi- University of Dayton

Books
Stephen Allan- Utah State University
Naeem Abdurrahman - University of Texas Austin
Anil Bajaj- Purdue University
Grant Baker - University of Alaska - Anchorage
Betty Barr - University of Houston

William Beckwith - Clemson University
Haym Benaroya- Rutgers University
Tom Bledsaw- ITT Technical Institute
Tom Bryson- University of Missouri, Rolla
Ramzi Bualuan - University of Notre Dame
Dan Budny- Purdue University
Dale Calkins - University of Washington
Arthur Clausing - University of Illinois
James Devine- University of South Florida
Patrick Fitzhorn - Colorado State University
Dale Elifrits- University of Missouri, Rolla
Frank Gerlitz - Washtenaw College
John Glover - University of Houston
John Graham - University of North Carolina-Charlotte
Malcom Heimer - Florida International University
A.S. Hodel - Auburn University
Vern Johnson- University of Arizona
Kathleen Kitto - Western Washington University
Robert Montgomery- Purdue University
Mark Nagurka- Marquette University
Ramarathnam Narasimhan- University of Miami
Larry Richards - University of Virginia
Marc H. Richman - Brown University
Avi Singhal-Arizona State University
Tim Sykes- Houston Community College
Thomas Hill- SUNY at Buffalo
Michael S. Wells - Tennessee Tech University
Joseph Wujek - University of California - Berkeley
Edward Young- University of South Carolina
Mandochehr Zoghi - University of Dayton

1

An Introduction to Engineering Problem Solving

GRAND CHALLENGE: WEATHER PREDICTION

Weather satellites provide a great deal of information to meteorologists who attempt to predict the weather. Large volumes of historical weather data can also be analyzed and used to test models for predicting weather. In general, we can do a reasonably good job of predicting overall weather patterns, but local weather phenomena, such as tornadoes, water spouts, and microbursts, are still difficult to forecast. Even predicting heavy rainfall or large hail from thunderstorms is not easy. Although Doppler radar helps to locate regions within storms that could contain tornadoes or microbursts, it only detects the events as they occur. This affords little time to issue warnings to populated areas or aircraft. Accurate and timely prediction of weather and associated weather phenomena remains an elusive goal.

1.1 GRAND CHALLENGES

Engineers solve real-world problems using scientific principles from disciplines that include computer science, mathematics, physics, and chemistry. It is this variety of subjects, and the challenge of real-world problems, that make engineering so interesting and so rewarding.

This section takes a look at a set of real-world problems. These "grand challenges", fundamental problems in science and engineering with broad potential impact, were identified by the Office of Science and Technology Policy in

SECTIONS

OBJECTIVES

In this chapter, you will

- First encounter a set of "grand challenges," such as weather prediction, that this text will use as examples.
- Read about the fundamentals of computer hardware and software.
- Learn a problem-solving methodology to describe a problem and develop its solution.

Washington, D.C. The grand challenges are part of a research and development strategy for high-performance computing. Solving these problems will require technological breakthroughs in both engineering and science. Just as computers played an important part in the engineering achievements of the last 35 years, computers will play an even greater role in solving problems of this scope and importance. The grand challenges described here will be encountered again later in the book.

The **prediction of weather, climate, and global change** requires an understanding of the coupled atmosphere and ocean biosphere system. This includes understanding CO_2 dynamics in the atmosphere and ocean, ozone depletion, and climatological changes due to the releases of chemicals or energy. This complex interaction also includes solar interactions. A major eruption from a solar storm near a "coronal hole" (a venting point for the solar wind) can eject vast amounts of hot gases from the sun's surface toward the earth's surface at speeds of over a million miles per hour. This ejection of hot gases bombards the earth with x-rays and can interfere with communication and cause power fluctuations in power lines. Learning to predict changes in weather, climate, and global change involves collecting large amounts of data for study and developing new mathematical models that can represent the interdependency of many variables.

Computerized speech understanding could revolutionize our communication systems, but many problems are involved. Systems are currently in use that "teach" a computer to understand words from a small vocabulary spoken by the same person. However, to develop systems that are speaker-independent and that understand words from large vocabularies and from different languages is difficult. Subtle changes in one's voice, such as those caused by a cold or stress, can affect the performance of speech recognition systems. And even when a computer does correctly recognize words, it is not simple to determine their meaning. Words are often context-dependent and cannot be analyzed separately. Intonation, such as raising one's voice, can change a statement into a question. Although many difficult problems remain in automatic speech recognition and understanding, exciting possible applications are everywhere. Imagine a telephone system that determines the languages being spoken and translates the speech signals so that each person hears the conversation in his or her native language.

The goal of the **Human Genome Project** is to locate, identify, and determine the function of each of the 50,000 to 100,000 genes contained in human DNA (deoxyribonucleic acid), which is the genetic material found in cells. The deciphering of the human genetic code will lead to many technical advances, including the ability to detect many of the over 4,000 known human genetic diseases, such as sickle-cell anemia and cystic fibrosis. However, deciphering the code is complicated by the nature of genetic information. Each gene is a double-helix strand composed of base pairs (adenine bonded with thymine or cytosine bonded with guanine) arranged in a step-like manner with phosphate groups along the side. These base pairs can occur in any sequential order, and represent the hereditary information in the gene. The number of base pairs in human DNA has been estimated to be around 3 billion. Because DNA directs the production of proteins for all metabolic needs, the proteins produced by a cell may provide a key to the sequence of base pairs in the DNA.

Substantial **improvements in vehicle performance** require more complex physical modeling in the areas of fluid dynamic behavior for three-dimensional flow fields and flow inside engine turbomachinery and ducts. Turbulence in fluid flows affects the stability and control, thermal characteristics, and fuel performance of aerospace vehicles. Modeling this flow is necessary for the analysis of new configurations. The analysis of the aeroelastic behavior of vehicles also affects new designs. So does the efficiency of combustion systems, because improving combustion efficiency requires understanding the relationships between the flows of the various substances and the chemistry that causes these substances to react. Vehicle performance is also being addressed through the use of onboard computers and microprocessors. Some cars already have transportation systems with small video screens mounted on the dash. The driver enters the destination, and the video screen shows the street names and route from the current to the desired location. A communication network keeps the car's computer aware of any traffic jams, so that it can automatically display a new route if necessary. Transportation research addresses totally automated driving systems, with computers and networks handling all the control and information interchange.

Enhanced oil and gas recovery will allow us to locate the estimated 300 billion barrels of oil reserves in the United States alone. Current methods use seismic techniques that can evaluate structures 20,000 feet below the surface that are likely to contain oil and gas. These techniques use a group of sensors (called a sensor array) located near the area to be tested. A ground shock signal is sent into the earth, reflected by the different geological layer boundaries, and received by the sensors. Using sophisticated signal processing, the boundary layers can be mapped and some estimate can be made as to the materials in the various layers, such as sandstone, shale, and water. The ground shock signals can be generated in several ways. A hole can be drilled, and a charge can be exploded in the hole. An explosive charge can also be set off on the surface. Or a special truck fitted with a hydraulic hammer can pound the earth several times per second. Continued research is needed to improve resolution of the information, and to find production and recovery methods that are economical and ecologically sound.

These grand challenges are only a few of the interesting problems waiting for engineers and scientists to solve. Solving problems of this magnitude requires organized approaches that combine ideas and technologies. Computers will play a key role.

1.2 COMPUTING SYSTEMS

Before discussing the C++ computer language, let's take a brief look at computers themselves. A **computer** is a machine designed to perform operations that are specified with programs. A **program** is a set of instructions that describe the steps the computer will perform.

Computer **hardware** refers to the computer's physical equipment, such as the keyboard, mouse, and printer. Computer **software** refers to the programs.

Computer Hardware

All computers have a common internal organization, as shown in Figure 1.1. The **processor** is the part of the computer that controls the other parts. It accepts input values (from a device such as a keyboard) and stores them in memory. It also interprets the instructions from a computer program.

In order to add two values, the processor will retrieve the values from memory and send them to the **arithmetic logic unit,** or ALU. The ALU performs the addition, and the processor then stores the result in memory. The processing unit and the ALU use **internal memory,** which is composed of read-only memory (ROM) and random access memory (RAM), in their processing. Most data are stored in external or secondary memory using hard disk drives or floppy disk drives that are attached to the processor.

The processor and ALU together are called the **central processing unit,** or CPU. A **microprocessor** is a CPU contained in a single integrated circuit chip. The millions of components of a microprocessor occupy an area smaller than a postage stamp.

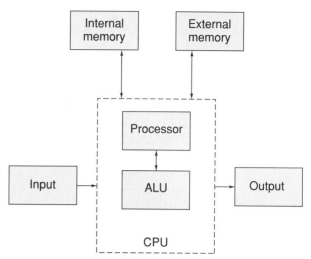

Figure 1.1.   Internal organization of a computer.

You usually tell the computer to display the values that it has computed on the terminal screen or to print them on paper using a printer. Laser printers use light beams to transfer images to paper. Dot matrix printers use a matrix (or grid) of pins to produce the shape of each character on paper. A computer can also write information to disks, which store the information magnetically. A printed copy of information is called a hard copy. A magnetic copy of information is called an electronic copy or a soft copy.

Computers come in many sizes, shapes, and forms. **Personal computers (PCs)** are small, fairly inexpensive computers commonly used in offices, homes, and laboratories. PCs are also referred to as microcomputers. Their design is built around a microprocessor, such as the Intel 486 microprocessor, which can process millions of instructions per second (mips). Minicomputers are more powerful than microcomputers. Mainframes are even more powerful, and are often used in businesses and research laboratories. A **workstation** is a minicomputer or mainframe computer small enough to fit on a desktop. **Supercomputers** are the fastest of all computers and can process billions of instructions per second. As a result of their speed, supercomputers are capable of solving very complex problems that cannot be feasibly solved on other computers. Mainframes and supercomputers require special facilities and a specialized staff to run and maintain the computer systems.

The type of computer needed to solve a particular problem depends on the problem requirements. If the computer is part of a home security system, a PC is sufficient. If the computer is running a flight simulator, a mainframe is probably needed. Computer **networks** allow computers to communicate with each other so that they can share resources and information. For example, ethernet is a commonly used local area network (LAN).

Computer Software

Computer software contains the instructions or commands that you want the computer to perform. Important categories of software include operating systems, software tools, and language compilers. Figure 1.2 illustrates the interaction between these categories of software and the computer hardware. Let's discuss each of these software categories in more detail.

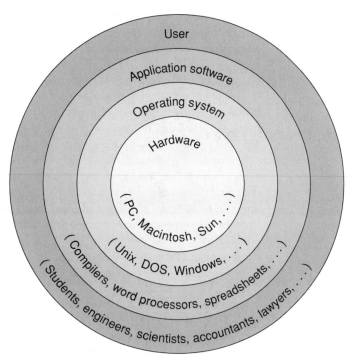

Figure 1.2.   Software interface to the computer.

Operating Systems   An **operating system** provides an interface between you (the user) and the hardware by providing a convenient and efficient environment in which you can select and run the other types of software on your system. An operating system typically comes with the computer hardware when you buy it.

Operating systems also contain a group of programs, called utilities, that you can use to perform tasks common to computer systems. These include printing files, copying files from one disk to another, and listing the files that you have saved on a disk. The commands that you use to carry out such tasks vary from computer to computer. For example, to list your files using DOS (a disk operating system used mainly with PCs), the command is **dir.** To list your files with UNIX (a powerful operating system frequently used with workstations), the command is **ls**. Some operating systems are easy to work with, presenting graphical alternatives on the screen for you to choose from rather than you having to type commands. Windows and the Macintosh environment are examples of user-friendly operating systems. C++ programs can be run on many types of computers, and a particular computer can also use different operating systems. It is not feasible to discuss the wide variety of operating systems that you might use while taking this course. Your professor will provide the specific operating system information that you need to use the computers available at your university. You can also find this information in the operating system manuals.

Software Tools   Programs that have been written to perform common operations are called **software tools.** For example, word processors, such as Microsoft Word and WordPerfect, are programs that help you enter and format text. With word processors you can move sentences and paragraphs, enter mathematical equations, and check your spelling and grammar. Word processors are also used to enter computer programs and store them in files. With the most sophisticated word processors, you can produce well-designed pages that combine elaborate charts and graphics with text and headlines. These programs use a technology called desktop publishing, which combines a powerful word processor with a high-quality printer to produce professional-looking documents.

Spreadsheet programs are software tools that allow you to work easily with data that can be displayed in a grid of rows and columns. Spreadsheets were initially used for financial and accounting applications, but many science and engineering problems can be solved easily using spreadsheets. Most spreadsheet packages include plotting capabilities, so they can be especially useful in analyzing and displaying information. Lotus 1-2-3®, Quattro Pro®, and Excel® are popular spreadsheet packages.

Another popular group of software tools are database management programs, such as dBASE IV® and Paradox®. With these programs you can store a large amount of data, and then easily retrieve pieces of the data and format them into reports. Databases are used by virtually all large and many small organizations, ranging from banks and airlines to the corner grocery store. Scientific databases are also used to analyze large amounts of data. Meteorology data is an example of scientific data that require large databases for storage and analysis.

Computer-aided design (CAD) packages, such as AutoCAD®, AutoSketch®, and CADKEY®, allow make it possible for you to define objects and then manipulate them graphically. For example, you can design an object and then view it from different angles or observe a rotation of the object from one position to another.

There are also some very powerful mathematical computation tools, such as MATLAB®, Mathematica®, Mathcad®, and Maple®. Not only do these tools have powerful mathematical commands, but they also provide extensive capabilities for generating graphs. This combination of computational power and visualization power makes them particularly useful tools for engineers.

If an engineering problem can be solved using a software tool, it is usually more efficient to use the existing tool than to write a program in a computer language to solve the problem. However, many problems cannot be solved using software tools; or a software tool may not be available on the computer system that must be used for solving the problem. Thus it is also necessary to know how to write programs using computer languages. The distinction between a software tool and a computer language is becoming less clear, as some of the more powerful tools, such as MATLAB and Mathematica, include their own language in addition to specialized operations.

Computer Languages   Computer languages can be described in terms of levels. **Low-level languages** or machine languages are the most primitive languages. **Machine language** is tied closely to the design of the computer hardware. Because computer designs are based on two-state technology (devices with two states, such as open or closed circuits, on or off switches, positive or negative charges), machine language is written using two symbols, which are usually represented using the digits 0 and 1. Therefore, machine language is also a binary language, and the instructions are written as sequences of 0's and 1's called binary strings. Because machine language is closely tied to the design of the computer hardware, the machine language for a Sun computer is different from the machine language for a Silicon Graphics computer.

An **assembly language** is also unique to a specific computer design, but its instructions are written in English-like statements instead of binary. Assembly languages usually do not have very many statements, and so writing programs can be tedious. In addition, to use an assembly language, you must also know information that relates to the specific computer hardware. Instrumentation that contains microprocessors often requires that programs, called **real-time programs**, operate very fast. These real-time programs are usually written in assembly language to take advantage of the specific computer hardware in order to perform the steps faster.

High-level languages are computer languages that have English-like commands and instructions. Examples of such languages are C, C++, Fortran, Ada, Pascal, COBOL, and Basic. Writing programs in high-level languages is certainly easier than writing programs in machine language or in assembly language. However, a high-level language contains a large number of commands and an extensive set of syntax (or grammar) rules for using the commands. To illustrate the syntax and punctuation required by both software tools and high-level languages, Table 1-1 shows how to compute the area of a circle with a specified diameter using several different languages and tools. Notice both the similarities and the differences in this simple computation.

Although C++ was classified here as a high-level language, many people like to describe C++ as a mid-level language, because it allows access to low-level routines and is often used to define programs that are converted to assembly language. Many software tools are written in C++.

Languages are also defined in terms of historical "generations." The first generation of computer languages to be developed was machine language, the second generation was assembly language, and the third was high-level languages. Fourth generation languages, also referred to as 4GLs, have not been developed yet and are described only in terms of characteristics and programmer productivity. The fifth generation of languages is called natural language. To program in a fifth generation language, one would use the syntax of natural speech. This requires the achievement of one of the grand challenges—computerized speech understanding.

Fortran (FORmula TRANslation) is a high-level computer language developed in the mid-1950s for solving engineering and scientific problems. New standards updated the language over the years. The current standard, Fortran 90, contains strong numerical computation capabilities, along with many of the new features and structures in languages such as C++. **COBOL** (COmmon Business-Oriented Language) was developed in the late 1950s to solve business problems.

Basic (Beginner's All-purpose Symbolic Instruction Code) was developed in the mid-1960s and was used as an educational tool. It is often included with the system software for a PC. **Pascal** was developed in the early 1970s and is widely used in computer science programs to introduce students to computing. **Ada** was developed at the initiative of the U.S. Department of Defense with the purpose of developing a language appropriate to embedded computer systems, which are typically implemented using microprocessors. The final design of the language was accepted in 1979; it was named in honor of Ada Lovelace, who developed instructions for doing computations on an analytical machine in the early 1800s.

TABLE 1-1   Comparison of Software Statements

SOFTWARE	EXAMPLE STATEMENT
C	`area = 3.141593*(diameter/2)*(diameter/2);`
MATLAB	`area = pi*((diameter/2)^2);`
Fortran	`area = 3.141593*(diameter/2.0)**2`
Ada	`area := 3.141593*(diameter/2)**2;`
Pascal	`area := 3.141593*(diameter/2)*(diameter/2)`
Basic	`let a = 3.141593*(d/2)*(d/2)`
COBOL	`compute area = 3.141593*(diameter/2)*(diameter/2).`

C is a general-purpose language that evolved from two languages, BCPL and B, that were developed at Bell Laboratories in the late 1960s. In 1972, Dennis Ritchie developed and implemented the first C compiler on a DEC PDP-11 computer at Bell Laboratories. The language became very popular for system development because it was hardware independent. Because of its popularity in both industry and in academia, it became clear that a standard definition was needed. A committee of the American National Standards Institute (ANSI) was created in 1983 to provide a machine-independent and unambiguous definition of C++. In 1989, the **ANSI C** standard was approved; that language is described in this text.

C++ has become the language of choice of many engineers and scientists because it has powerful commands and data structures and yet can be used easily for system-level operations. Because C++ is the language that a new engineer is most likely to encounter in a job, it is a good choice for an introduction to computing for engineers. However, in an introductory course, it is more important to establish a good foundation in computing than it is to cover all the features of the language. Therefore, the most important features of C++ for solving engineering problems are incorporated here, without attempting to cover all elements of the language.

Executing a Computer Program   A program written in a high-level language such as C++ must be translated into machine language before a computer can carry out, or **execute**, the instructions. A special program called a **compiler** performs this translation. Thus, in order to be able to write and execute C++ programs on a computer, the computer's software must include a C++ compiler. C++ compilers are available for the entire range of computer hardware, from supercomputers to personal computers. Most C++ compilers are based on the ANSI standards, but you should check the documentation of your compiler to see if it is an ANSI C++ compiler. If it is not an ANSI C++ compiler, there will be some differences between the C++ language discussed in this text and the C++ language accepted by the compiler.

If a compiler detects any errors (often called **bugs**) during compilation, error messages are printed. You must correct the program statements and then perform the compilation step again. The errors identified during this stage are called **compile errors** or compile-time errors. For example, to divide the value stored in a variable called sum by 3, the correct expression in C++ is sum/3. If you incorrectly write the expression using the backslash, as in sum\3, you will get a compiler error. The process of compiling, correcting statements (or **debugging**), and recompiling must often be repeated several times before the program compiles without compiler errors.

When there are no compiler errors, the compiler generates a program in machine language that performs the steps specified by the original C++ program. The original C++ program is referred to as the **source program**, and the machine language version is called an **object program.** Thus, the source program and the object program specify the same steps, but the source program is specified in a high-level language, and the **object program** is specified in machine language.

Once the program has compiled correctly, additional steps are necessary to prepare the object program for execution. This preparation involves **linking** other machine language statements to the object program, and then **loading** the program into memory. After this linking/loading, the program steps are executed by the computer. New errors, called execution errors, run-time errors, or **logic errors**, may be identified in this stage. They are also called program bugs. Execution errors often cause termination of a program. For example, the **program** statements may attempt to perform a division by zero, which generates an execution error. Some execution errors do not stop a program from executing, but they cause incorrect results to be computed. These types of errors can be caused by programmer mistakes in determining the correct steps in the solutions and by errors in the data processed by the program.

When execution errors occur due to errors in the program statements, you must correct the errors in the source program and then begin again with the compilation step. Even when a program appears to execute properly, check the answers carefully to be sure that they are correct. The computer will perform the steps precisely as specified, so if you specify the wrong steps, the computer will execute these wrong (but syntactically legal) steps and present an incorrect answer.

The processes of compilation, linking/loading, and execution are outlined in Figure 1.3. The process of converting an assembly language program to binary is performed by an **assembler** program. The corresponding processes are called assembly, linking/loading, and execution.

A C++ compiler often has additional capabilities that provide a user-friendly environment to implement and test C++ programs. For example, some C++ environments contain text processors so that program files can be generated, compiled, and executed in the same software package, as opposed to using a separate word processor that requires the use of operating system commands to transfer back and forth between the word processor and the compiler. Many C++ programming environments include **debugger** programs, which are useful in identifying errors in a program. Debugger programs show the values stored in variables at different points in a program and let us step through the program line by line.

As C++ statements are described in this text, common errors associated with the statements or useful techniques for locating such errors will also be covered. These debugging aids are summarized at the end of each chapter.

Software Life Cycle   In 1955, the cost of a typical computer solution was estimated to be 15% for software development and 85% for associated computer hardware. Over the years, the cost of the hardware has decreased dramatically, whereas the cost of the software has increased. In 1985, it was estimated that these numbers had essentially switched, with 85% of the cost for software and 15% for hardware. With the majority of the cost of a computer solution residing in software development, a great deal of attention has been given to understanding the development of a software solution.

The development of a software project generally follows definite steps or cycles, which are collectively called the **software life cycle**. These steps typically include project definition, the detailed specification, coding and modular testing, integrated testing, and maintenance. Data indicate that the corresponding percentages of effort involved can be estimated as shown in Table 1-2. From these estimates, it is clear that **software maintenance** is a significant part of the cost of a software system. This maintenance includes

Figure 1.3.   Program compilation/linking/execution.

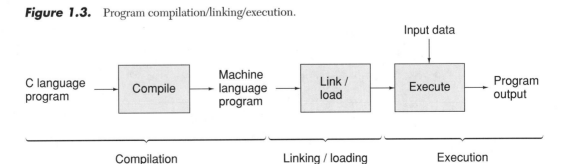

TABLE 1-2 Software Life Cycle Phases

LIFE CYCLE	PERCENT OF EFFORT
Definition	3%
Specification	15%
Coding and Modular testing	14%
Integrated Testing	8%
Maintenance	60%

adding enhancements to the software, fixing errors identified as the software is used, and adapting the software to work with new hardware and software. The ease of providing maintenance is directly related to the original definition and specification of the solution, because these steps lay the foundation for the rest of the project. The problem-solving process presented in the next section emphasizes the need to define and specify the solution carefully before beginning to code or test it.

A technique that has been successful in reducing the cost of software development both in time and in cost is the development of software prototypes. A **software prototype** is a simplified version of the final software, which can be tried out and modified early in the life cycle. A prototype does not contain all of the functions of the final software. But instead of waiting until the full-blown software system is developed and then letting users work with it and discover its shortcomings, a prototype can be changed with less time, effort, and expense.

As an engineer, it is very likely that you will need to modify or add additional capabilities to existing software. These modifications will be much simpler if the existing software is well-structured and readable and if the documentation that accompanies the software is up-to-date and clearly written. For these reasons, this text stresses developing good habits that make programs more readable and self-documenting. As new C++ statements are presented and new techniques are demonstrated, guidelines are included for writing well-structured and readable code.J

1.3 AN ENGINEERING PROBLEM-SOLVING METHODOLOGY

Problem solving is a key part of engineering courses as well as courses in computer science, mathematics, physics, and chemistry. Therefore, it is important to have a consistent approach to solving problems. It is also helpful if the approach is general enough to work for all these different areas, so that you do not have to learn one technique for mathematics problems, another for physics problems, and so on. The **problem-solving methodology** presented here works for engineering problems and can be tailored to solve problems in other areas as well.

The problem-solving method that will be used throughout this text has **five steps**:

1. State the problem clearly.
2. Describe the input and output information.
3. Work the problem by hand (or with a calculator) for a simple set of data.
4. Develop a solution and convert it to a computer program.
5. Test the solution with a variety of data.

Let's look at these steps individually, using an example of computing the distance between two points in a plane.

1. Problem Statement

The first step is to state the problem clearly. It is important to give a clear, concise problem statement to avoid any misunderstandings. For this example, here is the problem statement:

Compute the straight-line distance between two points in a plane.

2. Input/Output Description

The second step is to describe carefully the information that is given to solve the problem, and then identify the values to be computed. These items represent the input and the output for the problem (input/output or I/O). For many problems, a diagram that shows the input and output is useful. An **I/O diagram** shows the information that is used to compute the output, without defining the actual steps that will be used to determine the output. The I/O diagram for this example follows.

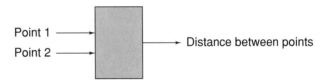

Point 1 ⟶
Point 2 ⟶ → Distance between points

3. Hand Example

The third step is to work the problem by hand or with a calculator, using a simple set of data. This is an important step and should not be skipped to save time, even for simple problems. This is the step in which you work out the details of the problem solution. If you cannot take a simple set of numbers and compute the output (either by hand or with a calculator), then you are not ready to move on to the next step. Instead, you should reread the problem, and perhaps consult reference material. The solution by hand for this specific example follows:

Let the points p_1 and p_2 have the following coordinates:

$$p_1 = 5 \ (1,5); \quad p_2 = 5 \ (4,7)$$

The distance between the two points is the hypotenuse of a right triangle, as shown in Figure 1.4. Using the Pythagorean theorem, you can compute the distance with the following equation:

$$\text{distance} = \sqrt{(side_1)^2 + (side_2)^2}$$

$$= \sqrt{(4-1)^2 + (7-5)^2}$$

$$= \sqrt{13}$$

$$= 3.61$$

4. Algorithm Development

Once you can work the problem for a simple set of data, you are then ready to develop an algorithm. An **algorithm** is a step-by-step outline of a problem solution. For simple problems such as this one, the algorithm can be listed as operations that are performed

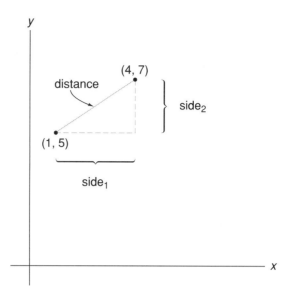

Figure 1.4.   Straight-line distance between two points.

one after another. This outline of steps decomposes the problem into simpler steps, as shown by the following outline of the steps required to compute and print the distance between two points.

Decomposition Outline

1. Give values to the two points.
2. Compute the lengths of the two sides of the right triangle generated by the two points.
3. Compute the distance between the two points, which is equal to the length of the hypotenuse of the triangle.
4. Print the distance between the two points.

This **decomposition outline** is then converted to C++ commands, so that you can use the computer to perform the computations. From the following solution you can see that the commands are very similar to the steps used in the hand example. The details of these commands will be explained in Chapter 2.

```
//-----------------------------------------------------------
//  Program chapter 1_1
//
//  This program computes the
//  distance between two points.

#include <iostream.h>
#include <stdlib.h>
#include <math.h>

int main()
{
   //  Define and initialize variables.
   double x1=1, y1=5, x2=4, y2=7,
          side_1, side_2, distance;
```

```
// Compute sides of a right triangle.
side_1 = x2 - x1;
side_2 = y2 - y1;
distance = sqrt(side_1*side_1 + side_2*side_2);

// Print distance.
cout << "The distance between the two points is "
     << distance << " cm." << endl;

// Exit program.
return EXIT_SUCCESS;
}
//-------------------------------------------------------
```

5. Testing

The final step in the problem-solving process is testing the solution. First test the solution with the data from the hand example, because you have already computed the solution. When the C++ statements in this solution are executed, the computer displays the following output:

```
The distance between the points is  3.61
```

This output matches the value calculated by hand. If the C++ solution does not match the hand solution, you need to review both solutions to find the error. Once the solution works for the hand example, test it with additional sets of data to be sure that the solution works for other valid sets of data.

KEY TERMS

algorithm	bug	execute
ANSI C	central processing unit (CPU)	external memory
arithmetic logic unit (ALU)	compiler	hardware
assembler	computer	high-level language
assembly language	debug	internal (secondary) memory
I/O diagram	personal computer (PC)	software prototype
low-level language	processor	software tool
machine language	program	source program
microprocessor	real-time program	supercomputer
network	software	workstation
object program	software life cycle	
operating system	software maintenance	

2

Simple C++ Programs

GRAND CHALLENGE: VEHICLE PERFORMANCE

Wind tunnels are test chambers built to generate precise wind speeds. Accurate scale models of new aircraft can be mounted on force-measuring supports in the test chamber. Measurements of the forces on the model can then be made at many different wind speeds and angles. Some wind tunnels operate at hypersonic velocities, generating wind speeds of thousands of miles per hour. The sizes of wind tunnel test sections vary from a few inches across to sizes large enough to accommodate a jet fighter. At the completion of a wind tunnel test series, many sets of data have been collected, which can be used to determine the lift, drag, and other aerodynamic performance characteristics of a new aircraft at its various operating speeds and positions.

OBJECTIVES

In this chapter, you will

- Learn to write a simple C++ program.
- Use C++ statements that define constants and variables.
- Learn how to do simple arithmetic operations in C++.
- Use C++ statements to read data from the keyboard, and to print information on the screen.
- Read about mathematical functions commonly used to solve engineering problems.

2.1 PROGRAM STRUCTURE

To see how a C++ program is structured, let's look again at the program introduced in Chapter 1. It computes and prints the distance between two points.

```
//------------------------------------------------------------
//  Program chapter 1_1
/ /
//  This program computes the
//  distance between two points.

#include <iostream.h>
#include <stdlib.h>
#include <math.h>

int main()
{
   //  Define and initialize variables.
  double x1=1, y1=5, x2=4, y2=7,
         side_1, side_2, distance;

   //  Compute sides of a right triangle.
  side_1 = x2 - x1;
  side_2 = y2 - y1;
  distance = sqrt(side_1*side_1 + side_2*side_2);

   //  Print distance.
  cout << "The distance between the two points is "
       << distance << " cm." << endl;

   //  Exit program.
  return EXIT_SUCCESS;
}
//------------------------------------------------------------
```

The first five lines of this program contain comments that give the program a name (chapter1_1) and define its purpose:

```
//------------------------------------------------------------
//  Program chapter 1_1
/ /
//  This program computes the
//  distance between two points.
```

There are two methods of inserting comments in a C++ program: A comment can begin with the characters //, or a comment can begin with the characters /* and end with the characters */. A comment can be on a line by itself, or it can be on the same line as a command. A comment can also extend over several lines. Each of the comment lines here is a separate comment because each line begins with /* and ends with */. Comments are optional, but good style requires that comments be used throughout a program to make it more readable both while writing the program and later when you or someone else needs to change the program. Use initial comments to give a name to the program and to describe the general purpose of the program. Include additional explanation comments throughout, to document what the program is doing. C++ allows comments and statements to begin anywhere on a line. Here the initial comments of a program start in the first column.

Preprocessor directives give instructions to the compiler that are performed before the program is compiled. The most common directive inserts additional statements in the program. This type of directive contains the characters #include followed by the name of the file that contains the additional statements. The example program contains the following three preprocessor directives:

```
#include <iostream.h>
#include <stdlib.h>
#include <math.h>
```

These directives specify that statements in the files `stdio.h`, `stdlib.h`, and `math.h` should be inserted in place of these three statements before the program is run. The `<` and `>` characters around file names indicate that the files are included with the **Standard C++ library.** This library is contained in the files that accompany an ANSI C++ compiler. The `stdio.h` file contains information related to the output statement used in this program, the `stdlib.h` file contains a constant that the program will use, and the `math.h` file contains information related to the square root function used in the program.

The h extension on these file names specifies that they are header files. Header files will be discussed in more detail later. Preprocessor directives are generally included after the initial comments describing the program's purpose.

Every C++ program contains a function named `main`. The body of the function is enclosed by braces, { }. In order to identify the body of the function easily, place these braces on lines by themselves. Thus, the two lines following the processor directives specify the beginning of the `main` function:

```
int main()
{
```

The `main` function contains two types of commands: declarations and statements. **Declarations** define the memory locations that statements will use, and therefore must precede the statements. The declarations may or may not give **initial values** to be stored in the memory locations. A comment precedes the declaration statement in this program:

```
//  Define and initialize variables.
double x1=1, y1=5, x2=4, y2=7,
       side_1, side_2, distance;
```

These declarations specify that the program will use seven variables named `x1`, `y1`, `x2`, `y2`, `side_1`, `side_2`, **and** `distance`. The term `double` indicates that the variables will store **double-precision floating-point** values. These variables can store values, such as 12.5 and 20.0005, with many digits of precision. In addition, this statement specifies that `x1` should be initialized (given an initial value) to the value 1, `y1` should be initialized to the value 5, `x2` should be initialized to the value 4, and `y2` should be initialized to the value 7. The initial values of `side_1`, `side_2`, **and** `distance` are not specified and should not be assumed to be initialized to zero. Because the declaration was too long for one line, the example splits it over two lines, with the second line indented to show that it is a continuation.

The **statements** that specify the operations to be performed in the example program are the following:

```
//  Compute sides of a right triangle.
side_1 = x2 - x1;
side_2 = y2 - y1;
distance = sqrt(side_1*side_1 + side_2*side_2);

//  Print distance.
cout << "The distance between the two points is "
     << distance << " cm." << endl;
```

These statements compute the lengths of the two sides of the right triangle formed by two points (see Figure 1.4), and then compute the length of the hypotenuse of the right triangle. The details of the syntax of these statements are discussed later in the chapter.

After computing the distance, it is printed with the `cout` statement. This output statement is too long for a single line, so it is separated into two lines, with the second line indented to show that it is a continuation. Additional comments explain the computations and the output statement. Also, note that the declarations and statements must end with a semicolon.

To exit the program, use a `return` statement. The constant `EXIT_SUCCESS` is defined in the `stdlib.h file`, and indicates a successful exit from the program.

```
//  Exit program.
return EXIT_SUCCESS;
```

Using a `return` statement at the end of the `main` function is optional in C++. But it is a good idea for documentation purposes.

The body of the `main` function ends with the right brace on a line by itself and another comment line to mark the end of the `main` function.

```
}
//------------------------------------------------------------
```

Note that there are some blank lines (also called white space) in the program, to separate the different components. These blank lines make a program more readable and easier to modify. The declarations and statements within the `main` function were indented three columns here in order to show the structure of the program. This kind of spacing provides a consistent style, and will make your programs easier to read.

Now that you have closely examined the C++ program from Chapter 1, let's compare its structure to the **general form** of a C++ program. The general form of a C++ program is:

```
introductory comments
preprocessing directives
int main()
{
    statements;
}
```

You will see this structure often in the programs to follow.

2.2 CONSTANTS AND VARIABLES

Constants and variables represent values that are used in programs. **Constants** are specific values, such as 2, 3.1416, or −1.5, that are included in C++ statements. **Variables** are memory locations that are given a name or **identifier**. The identifier is used to reference the value stored in the memory location. To understand memory locations and their corresponding identifiers, think of mailboxes with people's names on them. Each particular memory location (or mailbox) contain's a value (such as a letter or piece of junk mail). The following diagram shows the variables, their identifiers, and their initial values after the following declaration statement from program `chapter1_1`:

```
double x1=1, y1=5, x2=4, y2=7,
    side_1, side_2, distance;
```

x1	1	y1	5	x2	4
y2	7	side_1	?	side_2	?
distance	?				

The values of variables that were not given initial values are unspecified, and are indicated here with a question mark. Sometimes these values are called **garbage values**, because they are values left in memory from the previous program. A diagram that shows a variable along with its identifier and its value is called a **memory snapshot**, because it shows the contents of a memory location at a specific point in the execution of a program. The preceding memory snapshot shows the variables and their contents as specified by the declaration statement. When developing a program, it is a good idea to use memory snapshots to show the contents of variables both before and after a statement is executed, in order to show its effect.

Here are the rules for selecting a valid identifier:

- An identifier must begin with an alphabetic character or the underscore character _.
- Alphabetic characters in an identifier can be lowercase or uppercase.
- An identifier can contain digits, but not as the first character.
- An identifier can be of any length, but the first 31 characters of the identifier must be unique.

C++ is "case sensitive", which means that uppercase letters are different from lowercase letters. So Total, TOTAL, and total represent three different variables. Also, the variable distance_in_miles_from_earth_to_mars would not be distinguished from distance_in_miles_from_earth_to_venus, because the first 31 characters are the same. There are also some keywords with special meanings to the C++ compiler that you can't use for identifiers (Table 2-1).

Examples of valid identifiers are distance, x_1, X_Sum, average_measurement, and initial_time. Examples of invalid identifiers are 1x (begins with a digit), minimum-x (contains an invalid character —), I/O (contains an invalid character /), switch (a keyword), $sum (contains an invalid character $), and rate% (contains an invalid character %).

TABLE 2-1   Keywords

auto	double	int	struct
break	else	long	switch
case	enum	register	typedef
char	extern	return	union
const	float	short	unsigned
continue	for	signed void	volatile
default	goto	sizeof	while
do	if	static	

Carefully pick identifier names, so that they reflect the contents of the variables they identify. If possible, a name should also indicate the units of measurement. For example, if a variable represents a temperature measurement in degrees Fahrenheit, use an identifier such as `temp_F` or `degrees_F`. If a variable represents an angle, name it `theta_rad` to indicate that the angle is measured in radians, or `theta_deg` to indicate that the angle is measured in degrees.

PRACTICE!

Which of the following names are valid identifiers? If a name is not a valid identifier, give the reason that it is not acceptable, and suggest a replacement.

1. `xsum`
2. `x_sum`
3. `tax-rate`
4. `perimeter`
5. `sec^2`
6. `degrees_C`
7. `count`
8. `void`
9. `f(x)`
10. `m/s`
11. `Final_Value`
12. `w1.1`

2.2.1 Scientific Notation

Besides identifying variables used in a program, the declarations at the start of C++ functions must specify the types of values to be stored in the variables.

A **floating-point value** is one that can represent both integer and noninteger values, such as 2.5, −0.004, and 15.0. A floating-point value expressed in **scientific notation** is rewritten as a mantissa times a power of ten, where the **mantissa** has an absolute value greater than or equal to 1.0 and less than 10.0. For example, in scientific notation, 25.6 is written as 2.56×10^1, −0.004 is written as -4.0×10^{-3}, and 1.5 is written as 1.5×10^0.

In **exponential notation**, the letter e is used to separate the mantissa from the exponent of the power of ten. Thus, in exponential notation, 25.6 is written as 2.56e1, −0.004 is written as −4.0e−3, and 1.5 is written as 1.5e0.

The number of digits allowed by the computer for the decimal portion of the mantissa determines the **precision** or accuracy. The number of digits allowed for the exponent determines the **range**. Values with one digit of accuracy, and an exponent range of −8 to 7 could include values such as 2.3×10^5 (230,000) and 5.9×10^{-8} (0.000000059). This precision and exponent range would not be sufficient for many of the types of values used in engineering. For example, the distance in miles from Mars to the sun, with seven digits of precision, is 141,517,510 or 1.4151751×10^8; to represent this value we would need at least seven digits of accuracy and an exponent range that included the integer 8.

PRACTICE!

In problems 1–4, express the value in scientific notation. Specify the number of digits of precision needed to represent each value.

1. 35.004
2. 0.00042
3. −0.0999
4. 10,000,002.8

In problems 5–8, express the value in floating-point notation.

5. 1.03e−5
6. −1.05e5
7. −3.552e6
8. 6.67e−4

Numeric Data Types

Numeric data types specify the types of numbers that will be contained in variables. In C++ , numeric values are either integers or floating-point values (Figure 2.1). Nonnumeric data types (such as characters) are discussed in Chapter 6.

The **type specifiers** for signed integers are short, int, and long, which stand for short integer, integer, and long integer, respectively. The specific ranges of values are **system dependent**, which means that the ranges can vary from one system to another. On many systems, the short integer and the integer data types range from −32,768 to 32,767, and the long integer type often represents values from −2,147,483,648 to 2,147,483,647. (The unusual limits, such as 32,767 and 2,147,483,647, relate to conversions of binary values to decimal values.)

In C++ you can also add an unsigned qualifier to integer specifiers, where an unsigned integer represents only positive values. Signed and unsigned integers can represent the same number of values, but the ranges are different. For example, if an unsigned short has the range of values from 0 to 65,535, then a short integer has the range of values from −32,768 to 32,767; both variables can represent a total of 65,536 values.

Figure 2.1. Numeric data types.

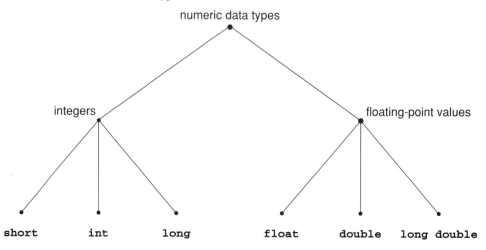

The **type specifiers** for floating-point values are `float` (single-precision), `double` (double-precision), and `long double` (extended precision). The following statement from program chapter1_1 thus defines seven variables that all contain double-precision floating-point values:

```
double x1=1, y1=5, x2=4, y2=7,
       side_1, side_2, distance;
```

The difference between the `float`, `double`, and `long double` types relates to the precision (or accuracy) and the range of the values represented. The precision and range are **system-dependent**. Table 2-2 contains precision and range information for integers and floating-point values used by the Borland C++ compiler, which can be used to run C programs as well as C++ programs. (C++ is an extension of C.) On most systems, a `double` data type stores about twice as many decimal digits of precision as are stored with a `float` data type. In addition, a `double` value will have a wider range of exponent values than a `float` value. The `long double` value may have more precision and a still wider exponent range, but this is again system dependent. A floating-point constant, such as 2.3, is assumed to be a `double` constant. To specify a `float` constant or a `long double` constant, the letter (or suffix) F or L must be appended to the constant. Thus, `2.3F` and `2.3L` represent a `float` constant and a `long double` constant, respectively.

Symbolic Constants

A **symbolic constant** is defined with a preprocessor directive that assigns an identifier to the constant. The directive can appear anywhere in a C++ program. The compiler will replace each occurrence of the directive identifier with the constant value in all statements that follow the directive. Engineering constants such as p or the acceleration of gravity are good candidates for symbolic constants. For example, consider the following preprocessing directive to assign the value 3.141593 to the variable PI with the following statement:

```
const double PI=3.141593;
```

TABLE 2-2 Example Data Type Limits*

INTEGERS	
`short`	maximum = 32,767
`int`	maximum = 32,767
`long`	maximum = 2,147,483,647
FLOATING POINT	
`float`	6 digits of precision
	maximum exponent = 38
	maximum value = 3.402823e + 38
`double`	15 digits of precision
	maximum exponent = 308
	maximum value = 1.797693e + 308
`long double`	19 digits of precision
	maximum exponent = 4932
	maximum value = 1.189731e + 4932

*Borland Turbo C++ 3.1 Compiler

Statements that need to use the value of π would then use the symbolic constant identifier instead of 3.141593, as illustrated in this statement

```
area = PI*radius*radius;
```

which computes the area of a circle.

Symbolic constants are usually defined with uppercase identifiers (as in PI instead of pi) to indicate that they are symbolic constants. Select the identifiers so that they are easy to remember. Finally, only one symbolic constant can be defined in a directive. Finally, several symbolic constants can be declared in one statement if they have the same data type. Const declarations end with a semicolon.

PRACTICE!

Give const declarations to assign symbolic constants for these constants.

1. speed of light, $c = 2.99792 \times 10^8$ m/s
2. charge of an electron, $e = 1.602177 \times 10^{-19}$ C
3. acceleration of gravity, $g = 9.8$ m/s^2
4. acceleration of gravity, $g = 32$ ft/s^2
5. radius of the moon, $r = 1.74 \times 10^6$ m

2.3 ASSIGNMENT STATEMENTS

Use the **assignment statement** to give a value to an identifier. The general form of the assignment statement is

```
identifier = expression;
```

where an **expression** can be a constant, another variable, or the result of an operation.

Consider the following two sets of statements that declare and give values to the variables sum and x1:

```
double sum=10.5;    double sum;
int x1=3;           int x1;
                      .
                      .
                      .
                    sum = 10.5;
                    x1 = 3;
```

After either set of statements is executed, the value of sum is 10.5, and the value of x1 is 3, as shown in the following memory snapshot.

```
sum         10.5            x1          3 3
```

The set of statements on the left define and initialize the variables at the same time. The assignment statements on the right could be used at any point in the program and thus may be used to change (as opposed to initialize) the values in variables. Note that a symbolic constant (such as PI, defined in the last section cannot be placed on the left side of an assignment statement because you cannot change its value in the program.

Multiple assignments are also allowed in C++ , as in the following statement, which assigns a value of zero to each of the variables x, y, and z:

```
x = y = z = 0;
```

Multiple assignments are discussed further at the end of this section.

You can also assign a value from one variable to another with an assignment statement:

```
rate = state_tax;
```

The equal sign should be read as "is assigned the value of." This statement then means "rate is assigned the value of `state_tax`." If `state_tax` contains the value 0.06, then rate also contains the value 0.06 after the statement is executed; the value in state_tax is not changed. Thus, the memory snapshots before and after this statement is executed are the following:

Before: rate | ? | state_tax | 0.06 |

After: rate | 0.06 | state_tax | 0.06 |

If you assign a value to a variable that has a different data type, then a conversion must occur during the execution of the statement. Sometimes the conversion can result in information being lost. For example, consider the following declaration and assignment statement:

```
int a;
   .
   .
   .
a = 12.8;
```

Because a is defined to be an integer, it cannot store a value with a nonzero decimal portion. Therefore, in this case, the memory snapshot after executing the assignment statement is the following:

a | 12 |

To determine if a **numeric conversion** will work properly, use the following order, which is from high to low:

high: long double
          double
          float
          long integer
          integer
low: short integer

If a value is moved to a data type that is higher in order, no information will be lost. But if a value is moved to a data type that is lower in order, information may be lost. Thus, moving an integer to a double will work properly, but moving a float to an integer may result in the loss of some information or in an incorrect result. In general, use only assignments that do not cause potential conversion problems. (Unsigned integers were not included in the list because errors can occur in both directions.) In general, use only assignments that will not cause conversion problems.

Arithmetic Operators

An assignment statement can be used to assign the result of an arithmetic operation to a variable. Here is an example, which computes the area of a square

```
area_square = side*side;
```

where * is used to indicate multiplication. The symbols + and – are used to indicate addition and subtraction, and the symbol / is used for division. Thus, each of the following statements is a valid computation for the area of a triangle:

```
area_triangle = 0.5*base*height;
area_triangle = (base*height)/2;
```

The use of parentheses in the second statement is not required but is used for readability.

Consider this assignment statement:

```
x = x + 1;
```

In algebra, this statement is invalid because a value cannot be equal to itself plus 1. However, this assignment statement should not be read as an equality. Instead, it means "x is assigned the value of x plus 1." With this interpretation, the statement indicates that the value stored in the variable x is incremented by 1. Thus, if the value of x is 5 before this statement is executed, then the value of x will be 6 after the statement is executed.

C++ also includes a **modulus** operator, %, which computes the remainder in a division between two integers. For example, 5%2 is equal to 1, 6%3 is equal to 0, and 2%7 is equal to 2. (The quotient of 2/7 is zero with a remainder of 2.) If a and b are integers, then the expression a/b computes the integer quotient, whereas the expression a%b computes the integer remainder. Thus, if a is equal to 9 and b is equal to 4, the value of a/b is 2, and the value of a%b is 1. An execution error occurs if the value of b is equal to zero in either a/b or a%b because the computer cannot perform **division by zero**. If either of the integer values in a and b is negative, the results of a/b and a%b are system dependent.

The modulus operator is useful in determining if an integer is a multiple of another number. For example, if a%2 is equal to zero, then a is even; otherwise a is odd. If a%5 is equal to zero, then a is a multiple of 5. The modulus operator is used a lot in developing engineering solutions.

The five operators (+, -, *, /, %) discussed in the previous paragraphs are **binary operators**—operators that operate on two values. C++ also includes **unary operators**—operators that operate on a single value. For example, plus and minus signs can be unary operators when they are used in an expression such as -x.

The result of a binary operation with values of the same type is another value of the same type. For example, if a and b are double values, then the result of a/b is also a double value. Similarly, if a and b are integers, then the result of a/b is also an integer. However, an integer division can sometimes produce unexpected results because any decimal portion of the integer division is dropped; the result is a **truncated result**, not a rounded result. Thus, 5/3 is equal to 1, and 3/6 is equal to 0.

An operation between values with different types is a **mixed operation**. Before the operation is performed, the value with the lower type is converted or promoted to the higher type, as discussed earlier. The operation is therefore performed with values of the same type. For example, if an operation is specified between an integer and a float, the integer will be converted to a float before the operation is performed, and the result will be a float.

Suppose that you want to compute the average of a set of integers. If the sum and the count of the integers have been stored in the integer variables sum and count, you might think that the following statements should correctly compute the average:

```
int sum, count;
float average;
  .
  .
  .
average = sum/count;
```

However, the division between these two integers gives an integer result that is then converted to a float value. Thus, if sum is 18 and count is 5, the value of average is 3.0, not 3.6. To compute this sum correctly, we use a **cast operator**—a unary operator that specifies a type change in a value before the next computation. In this example, the cast (float) is applied to sum:

```
average = (float)sum/(float)count;
```

The value of sum is converted to a float value before the division is performed. The division is then a mixed operation between a float value and an integer, so the value of count is converted to a float value. The result of the division is then a float value that is stored in average. If the value of sum is 18 and the value of count is 5, the value of average is now correctly computed to be 3.6. Note that the cast operator affects only the value used in the computation. It does not change the value stored in the variable sum.

PRACTICE!

Give the value computed by each of the following sets of statements.

```
1. int a = 27, b = 6, c;
   ...
   c = b%a;
```

```
2. int a = 27, b = 6;
   float c;
   ...
   c  =  a/(float)b;
```

```
3. int a;
   float b = 6, c  =  18.6;
   ...
   a  =  c/b;
```

```
4. int b = 6;
   float a, c = 18.6;
   ...
   a  =  (int)c/b;
```

Priority of Operators

In an expression that contains more than one arithmetic operator, the order in which the operations are performed is important. Table 2-3 shows the **precedence** of the arithmetic operators, which matches the standard precedence in algebra.

Operations within parentheses are always evaluated first. If parentheses are nested, the operations within the innermost parentheses are evaluated first. Unary operators are evaluated before the binary operators *, /, and %. Binary addition and subtraction are evaluated last. If there are several operators of the same precedence level in an expression, the variables or constants are grouped (or associated) with the operators in a specific order, as specified in Table 2-3. For example, consider the following expression:

```
a*b + b/c*d
```

Because multiplication and division have the same precedence level, and because the **associativity** (the order for grouping the operations) is from left to right, this expression will be evaluated as if it contained the following:

```
(a*b) + ((b/c)*d)
```

The precedence order does not specify whether a*b is evaluated before (b/c)*d; the order of evaluation of these terms is system dependent.

The spacing within an arithmetic expression is a style issue. Some people prefer to put spaces around each operator. A better idea might be to put spaces only around

TABLE 2-3 Precedence of Arithmetic Operators

PRECEDENCE	OPERATOR	ASSOCIATIVITY
1	parentheses: ( )	innermost first
2	unary operators: + – (type)	right to left
3	binary operators: * / %	left to right
4	binary operators: + –	left to right

binary addition and subtraction because they are evaluated last. Choose the spacing style that you prefer, and then use it consistently.

Assume that you want to compute the area of a trapezoid and that you have declared four double variables: `base`, `height_1`, `height_2`, and `area`. Assume further that the variables `base`, `height_1`, and `height_2` already have values. A statement to correctly compute the area of the trapezoid is

```
area = 0.5*base*(height_1 + height_2);
```

Suppose that you forgot the parentheses in the expression

```
area = 0.5*base*height_1 + height_2;
```

The statement would be executed as if it were this statement:

```
area = ((0.5*base)*height_1) + height_2;
```

Note that although you would get an incorrect answer, you would not get an error message to alert you to the error. Therefore, it is important to be very careful when converting expressions into C++ . In general, use parentheses to indicate the order of operations in a complicated expression to avoid confusion and to be sure that the expression is evaluated in the manner desired.

You may have noticed that there is not operator for an exponentiation operation to compute values such as x^4. A special mathematical function will be discussed later in the chapter to perform exponentiations. Of course, exponentiations with integer exponents, such as a^2, can be computed with repeated multiplications, as in `a*a`.

The evaluation of long expressions should be broken into several statements. For example, consider the following equation:

$$f = \frac{x^3 - 2x^2 + x - 6.3}{x^2 + 0.05005x - 3.14}$$

If you try to evaluate the expression in one statement, it becomes too long to be easily read:

```
f = (x*x*x - 2*x*x + x - 6.3)/(x*x + 0.05005*x - 3.14);
```

You could break the statement into two lines:

```
f = (x*x*x - 2*x*x + x - 6.3)/
    (x*x + 0.05005*x - 3.14);
```

Another solution is to compute the numerator and denominator separately:

```
numerator = x*x*x - 2*x*x + x - 6.3;
denominator = x*x + 0.05005*x - 3.14;
f = numerator/denominator;
```

The variables x, numerator, denominator, and f must be floating-point variables in order to compute the correct value of f.

PRACTICE!

In problems 1 and 2 give C++ statements to compute the indicated values. Assume that the identifiers in the expressions have been defined as `double` variables and have also been assigned appropriate values. Use the following constant:

acceleration of gravity: $g = 9.80665$ m/s²

1. Tension in a cord:

$$\text{tension} = \frac{2m_1 m_2}{m_1 + m_2} \cdot g$$

2. Fluid pressure at the end of a pipe:

$$P_2 = P_1 + \frac{\rho v_2^2 (A_2^2 - A_1^2)}{2A_1^2}$$

In problems 3 and 4, give the mathematical equations computed by the C++ statements. Assume that the following symbolic constants have been defined, where the units of G are m³/(kg · s²):

```
const double PI = 3.141593, G = 6.67259e-11;
```

3. Centripetal acceleration:
```
centripetal = 4*PI*PI*r/(T*T);
```

4. Change in potential energy:
```
change = G*M_E*m*(1/R_E  -  1/(R_E + h));
```

Overflow and Underflow

The values stored in a computer have a wide range of allowed values. However, if the result of a computation exceeds the range of allowed values, an error occurs. For example, assume that the exponent range of a floating point value is from −38 to 38. This range should accommodate most computations, but it is possible for the results of an expression to be outside of this range. Suppose that a program executes the following commands:

```
x = 2.5e30;
y = 1.0e30;
z = x*y;
```

The values of x and y are within the allowable range. However, the value of z should be 2.5e60, but this value exceeds the range. This error is called **exponent overflow**, because the exponent of the result of an arithmetic operation is too large to store in the memory assigned to the variable. The action generated by an exponent overflow is system dependent.

Exponent underflow is a similar error caused by the exponent of the result of an arithmetic operation being too small to store in the memory assigned to the variable. Using the same allowable range as in the previous example, the following commands give an exponent underflow:

```
x = 2.5e-30;
y = 1.0e30;
z = x/y;
```

Again, the values of x and y are within the allowable range, but the value of z should be 2.5e−60. Because the exponent is less than the minimum value allowed, an exponent underflow is caused. Again the action generated by an exponent underflow is system dependent. On some systems, the result of an operation with exponent underflow is set to zero.

Increment and Decrement Operators

The C++ language contains unary operators for incrementing and decrementing variables. You cannot use these operators with constants or expressions. The increment operator ++ and the decrement operator −− can be applied either in a prefix position (before the identifier) as in ++ count, or in a **postfix** position (after the identifier) as in count ++. If an increment or decrement operator is used by itself, it is equivalent to an assignment statement that increments or decrements the variable. Thus, the statement

```
y--;
```

is equal to this statement:

```
y = y - 1;
```

If the increment or decrement operator is used in an expression, the expression must be evaluated carefully. If the increment or decrement operator is in a prefix position, the identifier is modified, and the new value is used in evaluating the rest of the expression. If the increment or decrement operator is in a postfix position, the old value of the identifier is used to evaluate the rest of the expression, and then the identifier is modified. Thus, the execution of this statement

```
w = ++x - y; (2.1)
```
$$(2.1)$$

is equivalent to the execution of this pair of statements:

```
x = x + 1;
w = x - y;
```

Similarly, this statement

```
w = x++ - y; (2.2)
```
$$(2.2)$$

is equivalent to this pair of statements:

```
w = x - y;
x = x + 1;
```

When executing either (2.1) or (2.2), if you assume that the value of x is equal to 5 and the value of y is equal to 3, then the value of x increases to 6. However, after executing (2.1), the value of w is 3. After executing (2.2), the value of w is 2.

The increment and decrement operators have the same precedence as the other unary operators. If several unary operators are in an expression, they are associated from right to left.

Abbreviated Assignment Operators

C++ allows simple assignment statements to be abbreviated. For example, each pair of statements contains equivalent statements:

```
x = x + 3;
x += 3;
sum = sum + x;
sum += x;
```

```
d = d/4.5;
d /= 4.5;

r = r%2;
r %= 2;
```

In fact, any statement of this form

```
identifier = identifier operator expression;
```

can be written in this form:

```
identifier operator = expression;
```

Abbreviated assignment statements are usually used because they are shorter. Earlier in this section, the following **multiple-assignment** statement was used:

```
x = y = z = 0;
```

The interpretation of this statement is clear, but the interpretation of the following statement is not as evident:

```
a = b += c + d;
```

To evaluate this properly, use Table 2-4, which indicates that the assignment operators are evaluated last, and their associativity is right to left. Thus, the statement is equivalent to the following:

```
a = (b += (c + d));
```

Replacing the abbreviated forms with the longer forms of the operations gives

```
a = (b = b + (c + d));
```

or

```
b = b + (c + d);
a = b;
```

Evaluating this statement was good practice with the precedence/associativity table, but in general, statements used in a program should be more readable. Therefore, using abbreviated assignment statements in a multiple assignment statement is not recommended. Also, note that the spacing conventions used in this text insert spaces around abbreviated operators and multiple assignment operators, because these operators are evaluated after the arithmetic operators.

TABLE 2-4 Precedence of Arithmetic and Assignment Operators

PRECEDENCE	OPERATOR	ASSOCIATIVITY
1	parentheses: ( )	innermost first
2	unary operators: + - + + - - (type)	right to left
3	binary operators: * / %	left to right
4	binary operators: + -	left to right
5	assignment operators: = + = - = * = / = % =	right to left

PRACTICE!

Give a memory snapshot after each statement is executed, assuming that x is equal to 2 and that y is equal to 4 before the statement is executed. Also, assume that all the variables are integers.

1. z = x ++ * y;

2. z = ++ x * y;

3. x + = y;

4. y % = x;

2.4 STANDARD INPUT AND OUTPUT

So far, we've seen how to declare variables and then use them to compute new values. Now it's time to output the new values computed. In addition, a statement is described that lets us enter values from the keyboard when the program is run. To use either of these statements in a program, you must include this preprocessor directive:

```
#include <iostream.h>
```

This directive gives the compiler the information it needs to check references to the input/output functions in the Standard C++ library.

cout Statement

The cout statement allows us to print values and explanatory text to the screen. For example, consider the following statement, which prints the value of a variable named angle along with the corresponding units:

```
cout << "angle = " << angle << " radians" << endl;
```

The insertion operator << inserts information in the form of control strings (enclosed in double quotes) or variable expressions into an output stream of information that is is displayed on the screen when the endl (end line) reference is encountered. In this example, the << operators insert the information in the first control string into the output stream, inserts the value of the variable angle into the output stream, and insert the correct units into the output stream; then, the combined stream of information is displayed on the screen. If the value of angle is 2.841214, then the output generated by the statement is

```
angle = 2.841214 radians
```

In engineering, it is very important to include the corresponding units in the output along with the numerical values.

If a cout statement is long, you should split it into several lines, choosing a split that preserves readability. For example, a good split is generally before the operator and its accompanying control string or expression. To split a control string, break the text into two separate pieces of text, each in its own set of quotation marks. The following statements show several different ways to print the same output line:

```
cout << "distance between the points is "
     << distance << endl;

cout << "distance between the "
     <<  "points is " << distance << endl;

cout << "distance between the points is ";
cout << distance << endl;

cout << "distance between the ";
cout <<  "points is " << distance << endl;
```

Note the differences between splitting one long cout statement into several lines and rewriting the long cout statement as two short cout statements.

The backslash (\) is called an **escape character** when it is used in a control string. The compiler combines it with the character that follows it and then attaches a special meaning to the resulting combination of characters. The sequence \\ is used to insert a single backslash in a control string, and the sequence \" will insert a double quote in a control string. Thus, the output of the statement

```
cout << "\"The End.\"" << endl;
```

is

```
"The End."
```

The other **escape sequences** recognized by C++ are given in Table 2.5.

Formatted Output Functions

The Standard C++ library contains **format functions**—usually called **manipulators**—that can be used with the cout statement. For example, we can use the setw function to set the **field width** (the number of character positions that a number will occupy on the screen) for the next value that is displayed. The basic form of the setw manipulator is

```
setw(size)
```

TABLE 2-5 Escape Sequences

SEQUENCE	CHARACTER REPRESENTED
\a	alert (bell) character
\b	backspace
\f	form feed
\n	new line
\r	carriage return
\t	horizontal tab
\v	vertical tab
\\	backslash
\?	question mark
\'	single quote
\"	double quote

TABLE 2-6   Common Manipulators

MANIPULATOR	DESCRIPTION
setw(size)	sets the field width
setprecision(digit)	sets the precision
setfill(char)	fills the field with the character
ws	removes white spaces

This function contains one integer argument that specifies the number of characters to use in displaying the next value. For example, the following statement prints the value of x in a field of eight spaces:

```
cout << setw(8) << x;
```

The field width will be increased, if necessary, to more than eight spaces to print the value that is output. If the field width specifies more positions than are needed for the value, the value is right justified, which means that the extra positions to the left of the value are filled with blanks.

Note that the setw manipulator only specifies the field width for the next item in the cout statement. After that item is displayed, the field width reverts to the default value of 0. Therefore, the statement

```
cout << setw(10) << x << y;
```

only displays the value of x in a field that is 10 characters wide. The value of y, however, uses just enough space to display it completely.

The setprecision manipulator can be used to specify the precision of a value to be displayed (the number of difits after the decimal point). The basic form of the setprecision manipulator is

```
setprecision(digit)
```

where the argument of the function is an integer indicating the number of digits after the decimal point. For example, the statement

```
cout << setprecision(2) << setw(8) << x;
```

prints the value of x with the two digits after the decimal point, using a total field width of eight spaces. The decimal portion of a value is rounded to the specified precision; thus, the value 14.51578 of x will be printed as 14.52, with three blanks to its left. The setprecision manipulator is applied to all subsequent output.

The preprocessor directive

```
#include <iomanip.h>
```

must be inserted in a program if we plan to use a manipulator. Table 2.6 describes the commonly used manipulators.

PRACTICE!

Assume that the integer variable sum contains the value 150 and that the double variable average contains the value 12.368. Show the output line (or lines) generated by the following statements

```
1. cout << "sum = " << sum << endl << endl
        << "average = " << average << endl;
```

```
2. cout << "sum and average" << endl;
   cout << sum << setw(8) << setprecision(2)
           << average << endl;
3. cout << setprecision(2) << average
           << " is the average;" << endl;
   cout << sum << " is the sum" << endl;
4. cout << setprecision(2) << average
           << " is the average; ";
   cout << setw(6) << sum << " is the sum" << endl;
```

cin Statement

The cin statement allows us to enter values from the keyboard when a program is executed. For example, suppose that a program computes the number of acres of new forest growth after a specified period of time elapses. If the time elapsed is a constant in the program, we would have to change the value of the constant and then recompile and reexecute the program to obtain the output for a different period. Alternatively, if we use the cin statement to read the period, we do not need to recompile the program; we only need to reexecute it and enter the desired period from the keyboard.

For instance, if the value to be entered through the keyboard is an integer that is to be stored in the variable year, we could use the following cin statement to read the value:

```
cin >> year;
```

Here, >> is called the **input operator**. Since C++ supports the input of each standard data type (integer, floating point, and character), we can use the >> operator to obtain the input of any value from the keyboard. When we need multiple input values, the order of operations is simple: The first variable mentioned is input first, and the others are input in the order of their appearance, regardless of the data type involved. To illustrate, if we wish to read more than one value from the keyboard, we can use a cin statement with cascaded >> operators as follows:

```
int year;
double acres;
 .
 .
 .
cin >> acres >> year;
```

When the cin statement is executed, the program will read two values from the keyboard and convert them, in turn, into a double value and an int value. To help distinguish the input operator >> from the insertion operator <<, remember that the former directs the information toward the variable names for input and the latter directs the information toward the function name cout for output.

To prompt the program user to enter the input values, a cin statement is usually preceded by a cout statement that describes the information the user should enter from the keyboard:

```
cout << "Enter the acres of forest "
        "and the period (years):" << endl;
cin >> acres >> year;
```

The cout statement ends with a new-line specifier, so the values entered by the user will be on the line (or lines) following the prompt text. Thus, after the previous statements are executed and the user has responded to the prompt, the information on the screen might be

```
Enter the acres of forest and the period (years):
15.5 10
```

or

```
Enter the acres of forest and the period (years):
15.5
10
```

2.5 MATHEMATICAL FUNCTIONS

Arithmetic expressions that solve engineering problems often require computations other than addition, subtraction, multiplication, and division. For example, many expressions require the use of exponentiation, logarithms, exponentials, and trigonometric functions. This section discusses the mathematical functions available in the Standard C++ library. The following preprocessor directive should be used in programs using the mathematical functions:

```
#include <math.h>
```

This directive specifies that information be added to the program to aid the compiler when it converts references to the mathematical functions in the Standard C++ library. Before going over the rules relating to functions, let's look at a specific example. The following statement computes the sine of an angle theta and stores the result in the variable b:

```
b = sin(theta);
```

The sin function assumes that the argument is in radians. If the variable theta contains a value in degrees, you can convert the degrees to radians with a separate statement. (Recall that $180° = \pi$ radians.)

```
const double PI=3.141593;
   .
   .
   .
theta_rad = theta*PI/180;
b = sin(theta_rad);
```

The conversion can also be specified within the function reference:

```
b = sin(theta*PI/180);
```

Performing the conversion with a separate statement is usually preferable because it is easier to understand.

A function reference, such as sin(theta), represents a single value. The parentheses following the function name contain the inputs to the function, which are called **parameters** or **arguments**. A function may contain no arguments, one argument, or many arguments, depending on its definition. If a function contains more than one argument, it is important to list the arguments in the correct order. Some functions also require that the arguments be in specific units. For example, the trigonometric functions assume that arguments are in radians. Most of the mathematical functions assume that the arguments are double values; if a different type argument is used, it is converted to a `double` before the function is executed.

A function reference can also be part of the argument of another function reference. For example, the following statement computes the logarithm of the absolute value of x:

```
b = log(fabs(x));
```

When one function is used to compute the argument of another function, be sure to enclose the argument of each function in its own set of parentheses. This nesting of functions is also called composition of functions.

Now let's take a look at some functions commonly used in engineering computations. Other functions will be presented later in the book where relevant subjects are discussed. Tables of common functions are included on the last two pages of this book.

Elementary Math Functions

The elementary math functions include functions to perform a number of common computations, such as computing the absolute value of a number and the square root of a number. In addition, they also include a group of functions used to perform rounding. These functions assume that the type of each argument is `double`, and the functions all return a `double`. If an argument is not a double, a conversion will occur using the rules described in Section 2.3. Here is a list of these functions:

`fabs(x)`	Computes the absolute value of x.
`sqrt(x)`	Computes the square root of x, where $x \geq 0$.
`pow(x,y)`	Used for exponentiation. Computes the value of x to the y power, or x^y. Errors occur if x = 0 and $y \leq 0$, or if x, 0 and y is not an integer.
`ceil(x)`	Rounds x to the nearest integer toward ∞ (infinity). For example, `ceil(2.01)` is equal to 3.
`floor(x)`	Rounds x to the nearest integer toward 2 ∞ (negative infinity). For example, `floor(2.01)` is equal to 2.
`exp(x)`	Computes the value of e^x, where e is the base for natural logarithms, or approximately 2.718282.
`log(x)`	Returns ln x, the natural logarithm of x to the base e. Errors occur if $x \leq 0$.
`log10(x)`	Returns $\log_{10}x$, the common logarithm of x to the base 10. Errors occur if $x \leq 0$.

Remember that the logarithm of a negative value or zero does not exist. Thus, an execution error occurs if you use a logarithm function with a negative or zero value for its argument.

An additional mathematical function that you may find useful is the abs function. This function computes the absolute value of an integer and returns an integer value.

The header file containing information relative to this function is stdlib.h, and it should be included in programs referencing this function.

PRACTICE!

Evaluate the following expressions:

1. `floor(-2.6)`

2. `ceil(-2.6)`

3. `pow(2, -3)`

4. `sqrt(floor(10.7))`

Trigonometric Functions

The trigonometric functions assume that all arguments are of type `double`, and they return values of type `double`. In addition, the trigonometric functions assume that angles are represented in radians. To convert radians to degrees, or degrees to radians, use the following conversions, which use the fact that $180° = \pi$ radians:

```
const double PI=3.141593;
        .
        .
        .
angle_deg = angle_rad*(180/PI);
angle_rad = angle_deg*(PI/180);
```

The trigonometric functions are included in the Standard C++ library. A preprocessor directive including the information in `math.h` should be used with these functions. A brief summary of the functions follows:

`sin(x)`	Computes the sine of x, where x is in radians.
`cos(x)`	Computes the cosine of x, where x is in radians.
`tan(x)`	Calculates the tangent of x, where x is in radians.
`asin(x)`	Computes the arcsine or inverse sine of x, where x must be in the range [−1, 1]. The function returns an angle in radians in the range $[-\pi/2, \pi/2]$.
`acos(x)`	Calculates the arccosine or inverse cosine of x, where x must be in the range [−1, 1]. The function returns an angle in radians in the range $[0, \pi]$.
`atan(x)`	Computes the arctangent or inverse tangent of x. The function returns an angle in radians in the range $[-\pi/2, \pi/2]$.
`atan2(y,x)`	Calculates the arctangent or inverse tangent of the value y/x. The function returns an angle in radians in the range $[-\pi, \pi]$.

Note that the `atan` function always returns an angle in Quadrant I or IV, whereas the `atan2` function returns an angle that can be in any quadrant, depending on the signs of x and y. Thus, in many applications, the `atan2` function is preferred over the `atan` function.

PRACTICE!

In problems 1 and 2, give assignment statements for computing the indicated values, assuming that the variables have been declared and given appropriate values. Also assume that the following declaration has been made:

1. Length contraction:

$$\text{length} = k\sqrt{1 - \left(\frac{v}{c}\right)^2}$$

2. Distance of the center of gravity from a reference plane in a hollow cylinder sector:

$$\text{center} = \frac{38.1972 \cdot (r^3 - s^3) \cdot \sin a}{(r^2 - s^2) \cdot a}$$

In problems 3 and 4, give the equations that correspond to the assignment statements.

3. Range for a projectile:

```
range = (v0*v0/g)*sin(2*theta);
```

4. Speed of a disk at the bottom of an incline:

```
v = sqrt(2*g*h/(1 + I/(m*pow(r,2))));
```

2.6 PROBLEM SOLVING APPLIED: VELOCITY COMPUTATION

This section performs computations in an application related to the vehicle performance grand challenge. An advanced turboprop engine called the **unducted fan (UDF)** is one of the promising new propulsion technologies being developed for future transport aircraft. Turboprop engines, in use for decades, combine the power and reliability of jet engines with the efficiency of propellers. They are a significant improvement over earlier piston-powered propeller engines. Their application has been limited to smaller commuter-type aircraft, however, because they are not as fast or powerful as the fanjet engines used on larger airliners. The UDF engine employs significant advancements in propeller technology, which narrow the performance gap between turboprops and fanjets. New materials, blade shapes, and higher rotation speeds enable UDF-powered aircraft to fly almost as fast as fanjets, and with greater fuel efficiency. The UDF is also significantly quieter than the conventional turboprop.

During a test flight of a UDF-powered aircraft, the test pilot has set the engine power level at 40,000 Newtons, which causes the 20,000–kg aircraft to attain a cruise speed of 180 m/s (meters/second). The engine throttles are then set to a power level of 60,000 Newtons, and the aircraft begins to accelerate. As the speed of the plane increases, the aerodynamic drag increases in proportion to the square of the airspeed. Eventually, the aircraft reaches a new cruise speed where the thrust from the UDF engines is just offset by the drag. The equations used to estimate the velocity and acceleration of the aircraft from the time that the throttle is reset until the plane reaches its new cruise speed (at approximately 120 s) are the following:

$$\text{velocity} = 0.00001 \text{ time}^3 - 0.00488 \text{ time}^2 + 0.75795 \text{ time} + 181.3566$$

$$\text{acceleration} = 3 - 0.000062 \text{ velocity}^2$$

Plots of these functions are shown in Figure 2.2. Note that the acceleration approaches zero as the velocity approaches its new cruise speed.

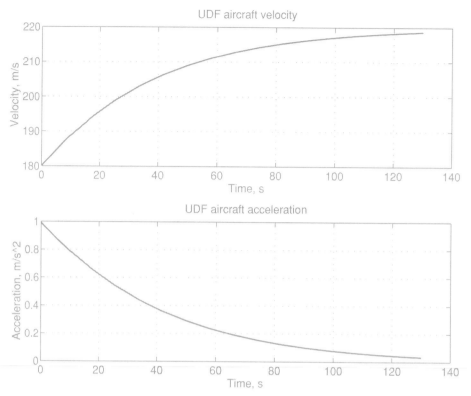

Figure 2.2. UDF aircraft velocity and acceleration.

Write a program that asks the user to enter a time value that represents the time elapsed (in seconds) since the power level was increased. Compute and print the corresponding acceleration and velocity of the aircraft at the new time value.

1. Problem Statement

Compute the new velocity and acceleration of the aircraft after a change in power level.

2.6.1 2. Input/Output Description

The following diagram shows that the input to the program is a time value and that the output of the program is the pair of new velocity and acceleration values.

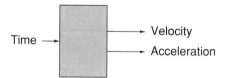

3. Hand Example

Suppose that the new time value is 50 seconds. Using the equations given for the velocity and accelerations, you can compute these values:

$$\text{velocity} = 208.3 \text{ m/s}$$
$$\text{acceleration} = 0.31 \text{ m/s2}$$

4. Algorithm Development

The first step in the development of an algorithm is the decomposition of the problem solution into a set of sequentially executed steps:

Decomposition Outline

1. Read new time value.
2. Compute corresponding velocity and acceleration values.
3. Print new velocity and acceleration.

Because this program is a simple program, the decomposition can be directly converted to C++ .

```cpp
//-------------------------------------------------------
//  Program chapter2_1
/ /
//  This program estimates new velocity and
//  acceleration values for a specified time.
#include <iostream.h>
#include <iomanip.h>
#include <stdlib.h>
#include <math.h>
int main()
{
   //  Define variables.
   double time, velocity, acceleration;
   //  Get time value from the keyboard.
   cout << "Enter new time value in seconds: ";
   cin >> time;
   //  Compute velocity and acceleration.
   velocity = 0.00001*pow(time,3) - 0.00488*pow(time,2)
            + 0.75795*time + 181.3566;
   acceleration = 3 - 0.000062*velocity*velocity;
   //  Print velocity and acceleration.
   cout << "velocity = " << setprecision(3)
        << velocity << " m/s" << endl;
   cout << "acceleration = " << setprecision(3)
        << acceleration << " m/s^2" << endl;
   //  Exit program.
   return EXIT_SUCCESS;
}
//-------------------------------------------------------
```

5. Testing

First test the program using the data from the hand example. This generates the following interaction:

```
Enter new time value in seconds: 50
velocity = 208.304 m/s
acceleration = 0.31 m/s^2
```

Because the values computed match the hand example, we can then test the program with other time values. If the values had not matched the hand example, we would need to determine if the error was in the hand example or in the program.

C++ STATEMENT SUMMARY

Preprocessor directives to include information from the files in the Standard C++ library:

```
#include <iostream.h>
#include <iomanip.h>
#include <stdlib.h>
#include <math.h>
```

Preprocessor directive to define a symbolic constant

```
const double PI=3.141593;
const int PEOPLE=50;
```

Declarations for integers

```
short sum=0;
int year_1, year_2;
long k;
```

Declarations for floating-point values

```
float height_1, height_2;
double length=10, side1, side2;
long double distance, velocity;
```

Assignment statement

```
area = 0.5*base*(height_1 + height_2);
```

Keyboard input statement

```
cin >> year;
```

Screen output statement

```
cout << "area is " << area
     << " square feet" << endl;
```

Program exit statement

```
return EXIT_SUCCESS;
```

Debugging Notes

1. Declarations and C++ statements must end with a semicolon.
2. Preprocessor directives do not end with a semicolon.
3. If possible, avoid assignments that could potentially cause information to be lost.
4. Use parentheses in a long expression to be sure that it is evaluated as desired.
5. Use double precision or extended precision to avoid problems with exponent overflow or underflow.
6. Be sure that the specifier matches the variable type in a `scanf` statement.
7. Errors can occur if user input values cannot be converted correctly to the specifier variable type in a `scanf` statement.
8. Do not forget the address operator with identifiers in the `scanf` statement.
9. Remember that symbolic constant definitions do not end with a semicolon.
10. In nested function references, each set of arguments must be in its own set of parentheses.

11. Remember that the logarithmic functions cannot be used with negative or zero values for arguments.

12. Be sure to use angles in radians with the trigonometric functions.

13. Remember that many of the inverse trigonometric functions and hyperbolic functions have restrictions on the ranges of allowable input values.

KEY TERMS

address operator	expression	prefix
argument	field width	preprocessor directive
assignment statement	floating-point value	prompt
binary operator	identifier	Standard C++ library
cast operator	keyword	symbolic constant
comment	memory snapshot	system dependent
econstant	multiple assignment	truncate
control string	overflow	unary operator
conversion specifier	parameter	underflow
declaration	postfix	variable
escape character	precedence	

Problems

Conversions. This set of problems involves conversions of a value in one unit to another unit. Each program should prompt the user for a value in the specified units and then print the converted value, along with the new units.

1. Write a program to convert miles to kilometers. (Recall that 1 mi $= 1.6093440$ km.)

2. Write a program to convert pounds to kilograms. (Recall that 1 kg $= 2.205$ lb.)

3. Write a program that converts degrees Fahrenheit (T_F) to degrees Rankin (T_R). (Recall that $T_F = T_R - 459.67°$.)

Areas and Volumes. These problems involve computing an area or a volume using input from the user. Each program should include a prompt to the user to enter the variables needed.

4. Write a program to compute the area of a triangle with base b and height h. (Recall that $A = \frac{1}{2}bh$.)

5. Write a program to compute the area of a sector of a circle when d is the angle in degrees between the radii. (Recall that $A = r^2\theta/2$, where θ is the angle in radians.)

6. Write a program to compute the volume of a sphere of radius r. (Recall that $V = 4/3\pi r^3$.)

Amino Acid Molecular Weights. The amino acids in proteins are composed of atoms of oxygen, carbon, nitrogen, sulfur, and hydrogen, as shown in Table 2-7. The molecular weights of the individual elements are

ELEMENT	ATOMIC WEIGHT
oxygen	15.9994
carbon	12.011
nitrogen	14.00674
sulfur	32.066
hydrogen	1.00794

TABLE 2-7 Amino Acid Molecules

AMINO ACID	O	C	N	S	H
Alanine	2	3	1	0	7
Arginine	2	6	4	0	15
Asparagine	3	4	2	0	8
Aspartic	4	4	1	0	6
Cysteine	2	3	1	1	7
Glutamic	4	5	1	0	8
Glutamine	3	5	2	0	10
Glycine	2	2	1	0	5
Histidine	2	6	3	0	10
Isoleucine	2	6	1	0	13
Leucine	2	6	1	0	13
Lysine	2	6	2	0	15
Methionine	2	5	1	1	11
Phenylanlanine	2	9	1	0	11
Proline	2	5	1	0	10
Serine	3	3	1	0	7
Threonine	3	4	1	0	9
Tryptophan	2	11	2	0	11
Tyrosine	3	9	1	0	11
Valine	2	5	1	0	11

7. Write a program that asks the user to enter the number of atoms of each of the five elements for an amino acid. Then compute and print the molecular weight for this amino acid.

8. Write a program that asks the user to enter the number of atoms of each of the five elements for an amino acid. Then compute and print the average weight of the atoms in the amino acid.

Logarithms to the Base b. To compute the logarithm of x to base b, we can use the following relationship:

$$\log_b x = \frac{\log_e x}{\log_e b}$$

9. Write a program that reads a positive number and then computes and prints the logarithm of the value to base 2. For example, the logarithm of 8 to the base 2 is 3, because $2^3 = 8$.

10. Write a program that reads a positive number and then computes and prints the logarithm of the value to the base 8. For example, the logarithm of 64 to the base 8 is 2, because $8^2 = 64$.

3

Control Structures and Data Files

Weather balloons are used to collect data from the upper atmosphere. The balloons are filled with helium. They rise to an equilibrium point, where the difference between the densities of the helium inside the balloon and the air outside the balloon is just enough to support the balloon's weight. During the day, the sun warms the balloon, causing it to rise to a new equilibrium point. In the evening, the balloon cools, and it descends to a lower altitude. The balloon can be used to measure the temperature, pressure, humidity, chemical concentrations, or other properties of the air nearby. A weather balloon may stay aloft for only a few hours, or as long as several years, collecting environmental data. As the helium leaks out or is released, the balloon falls back to earth.

SECTIONS

OBJECTIVES

In this chapter, you will

- Learn how to develop and refine algorithms.
- Become familiar with the general principles of structured programming.
- Program using sequence, selection, and loop structures.
- Learn the basics of reading and writing data files.

3.1 ALGORITHM DEVELOPMENT

In Chapter 2, the C++ programs shown were very simple. The steps were sequential. The steps typically involved reading data from the keyboard, computing new information, and then displaying it on the screen. In solving engineering problems, most of the solutions require additional steps. We need to expand the algorithm development part of the problem-solving process. An algorithm is a step-by-step outline of a problem solution.

Top-Down Design

Top-down design presents a "big picture" description of a problem solution in sequential steps. This overall description of the problem is then refined, until the steps become detailed enough to translate to programming language statements.

Decomposition Outline Shown in a diagram or written out in sequential text statements, a decomposition outline is the first definition of a problem solution. Decomposition outlines were already used in earlier chapters. For very simple problems, you can go from the decomposition outline directly to the C++ statements:

Decomposition Outline

1. Read new time value.
2. Compute corresponding velocity and acceleration values.
3. Print new velocity and acceleration.

For most problem solutions, you need to refine the decomposition outline into a description with more detail. This process is often referred to as a "divide-and-conquer strategy," because you keep breaking the problem solution into smaller and smaller portions. To describe this **stepwise refinement**, use pseudocode or flowcharts.

Refinement with Pseudocode and Flowcharts The refinement of an outline into more detailed steps can be done with pseudocode or a flowchart. **Pseudocode** uses English-like statements to describe the steps in an algorithm. A **flowchart** uses a diagram to describe the steps in an algorithm. The fundamental steps in most algorithms are shown in Figure 3.1, along with the corresponding notation in pseudocode and flowcharts.

Both pseudocode and flowcharts are commonly used tools, although both are generally not used with the same problem. Sometimes you need to go through several levels of pseudocode or flowcharts to develop complex problem solutions. This is the stepwise refinement mentioned earlier.

Decomposition outlines, pseudocode, and flowcharts are working models of a solution and are not unique. Each person working on a solution will have different decomposition outlines and pseudocode or flowchart descriptions, just as the C++ programs developed by different people will be somewhat different, although they solve the same problem. Keeping these differences between programs within certain boundaries—so that programs written by one person or group can more readily be understood and changed by another person or group—is the intent of structured programming.

Structured Programming

A **structured program** is one written using simple control structures to organize the solution to a problem. A simple control structure is usually defined to be sequence, selection, or repetition. The **sequence** structure contains steps that are performed one after another, in sequential order. The **selection** structure contains one set of steps that is performed if a condition is true, and another set of steps that is performed if the condition is false. The **repetition** structure contains a set of steps that is repeated as long as a condition is true. Let's look a little more closely at these control structures.

Sequence A sequence contains steps that are performed one after another. All the programs presented so far in the text have a sequence structure. For example, the flow-

Basic Operation	Pseudocode Notation	Flowchart Symbol

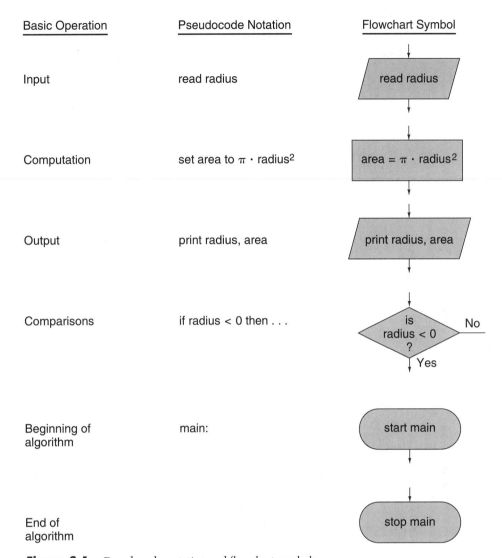

Input	read radius	read radius
Computation	set area to $\pi \cdot radius^2$	$area = \pi \cdot radius^2$
Output	print radius, area	print radius, area
Comparisons	if radius < 0 then . . .	is radius < 0 ?
Beginning of algorithm	main:	start main
End of algorithm		stop main

Figure 3.1.   Pseudocode notation and flowchart symbols.

chart for the program that computed the velocity and acceleration of the aircraft with the unducted engine is shown in Figure 3.2 on page 46.

Selection   A selection structure contains a condition that can be evaluated as either true or false. If the condition is true, one set of statements is executed. If the condition is false, another set of statements is executed.

For example, suppose that you have computed values for the numerator and denominator of a fraction. Before computing the division, however, you want to be sure that the denominator is not close to zero. Therefore, the condition that you want to test is "denominator close to zero." If the condition is true, you could print a message indicating that the value can't be computed. If the condition is false, which means that the denominator is not close to zero, then you want to compute and print the value of the fraction. In defining this condition, you also need to define "close to zero." For this example, let's assume that close to zero means that the absolute value is less than

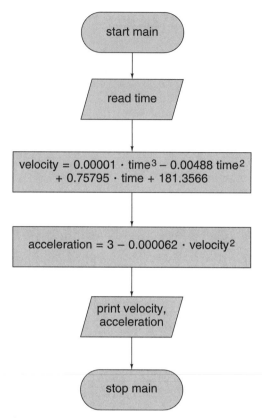

Figure 3.2.   Flowchart for inducted fan problem solution from Section 2.6.

0.0001. A pseudocode description follows, and a flowchart description of this structure is shown in Figure 3.3.

> *if* |*denominator*| < *0.0001*
>   *print "Denominator close to zero"*
> *else*
>   *set fraction to numerator/denominator*
>   *print fraction*

Note that this structure also contains a sequence structure (compute a fraction and then print the fraction), which is executed when the condition is false. You will see more variations of the selection structure later in the chapter.

Repetition   The repetition structure repeats a set of steps as long as a condition is true. For example, you might want to compute a set of velocity values that correspond to time values of 0, 1, 2, . . . , 10 seconds. It would be tedious to develop a sequential structure that has a statement to compute the velocity for a time of 0, another statement to compute the velocity for a time of 1, another to compute the velocity for a time of 2, and so on. Although this structure would require only 11 statements in this case, it could require hundreds or thousands of statements to compute the velocity values over a long period of time. To use the repetition structure, you could first initialize the time to 0. Then, as long as the time value is less than or equal to 10, you could compute and print a velocity value and increment the time value by 1. When the time value is greater than 10, exit the structure. Figure 3.4 contains the flowchart for this repetition structure. The pseudocode follows.

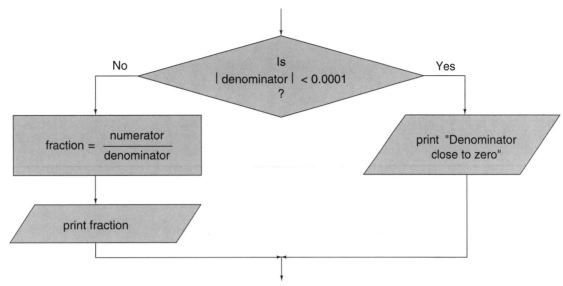

Figure 3.3. Flowchart for selection structure.

Figure 3.4. Flowchart for repetition structure.

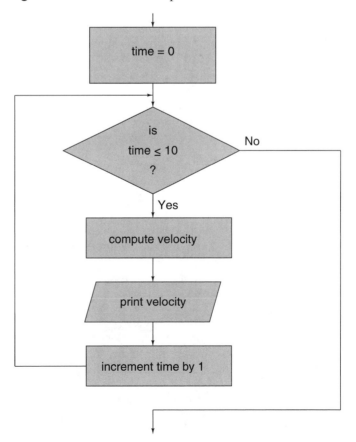

set time to 0
while time ≤ 10
    compute velocity
    print velocity
    increment time by 1

3.2 CONDITIONAL EXPRESSIONS

Both selection and repetition control structures use conditions. Therefore you need to better understand conditions before learning the C++ statements to use with the selection and repetition structures. A **condition** is an expression that can be evaluated as true or false, and is composed of expressions combined with relational or logical operators.

Relational Operators

Relational operators can be used to compare two expressions. The relational operators used in C++ are shown in the following list:

Relational Operator	Interpretation
<	less than
<=	less than or equal to
>	greater than
>=	greater than or equal to
==	equal to
!=	unequal to

Blanks can be used on either side of a relational operator, but blanks cannot be used to separate a two-character operator such as ==.

Here are some example conditions:

```
a < b
x+y >= 10.5
fabs (denominator) < 0.0001
```

Given the values of the identifiers in these conditions, you can evaluate each one to be true or false. For example, if a is equal to 5 and b is equal to 8.4, then a < b is a true condition. If x is equal to 2.3 and y is equal to 4.1, then x+y >= 10.5 is a false condition. If denominator is equal to 20.0025, then fabs(denominator)<0.0001 is a false condition.

Note that spaces are used around the relational operator in a logical expression, but not around the arithmetic operators in the conditions.

In C++, a true condition is assigned a value of 1, and, a false condition is assigned a value of zero. Therefore, the following statement is valid:

```
d = b>c;
```

If b>c, the value of d is 1; otherwise, the value of d is zero. Because a condition is given a value, it is valid to use a value in place of a condition. For example, consider the following statement:

```
if (a)
    count++;
```

If the condition value is zero, the condition is assumed to be false. If the condition value is nonzero, the condition is assumed to be true. Therefore, in the statement above, the value of count will be incremented if a is nonzero.

Logical Operators

Logical operators can also be used within conditions. However, logical operators compare *conditions*, not expressions. C++ supports three logical operators: not, and, or. These logical operators are represented by the following symbols:

LOGICAL OPERATOR	SYMBOL
not	!
and	&&
or	//

For example, consider the following condition:

```
a<b && b<c
```

The relational operators have higher precedence than the logical operator; therefore, this condition is read "a is less than b, and b is less than c." In order to make a logical statement more readable, insert spaces around the logical operator, but not around the relational operators. Given values for a, b, and c, you can evaluate this condition as true or false. For example, if a is equal to 1, b is equal to 5, and c is equal to 8, then the condition is true. If a is equal to −2, b is equal to 9, and c is equal to 2, then the condition is false.

If A and B are conditions, the logical operators can be used to generate new conditions A && B, A // B, !A, and !B. The condition A && B is true only if both A and B are true. The condition A // B is true if either or both A and B are true. The ! operator changes the value of the condition with which it is used. Thus, the condition !A is true only if A is false, and the condition !B is true only if B is false. These definitions are summarized in Table 3-1.

When expressions with logical operators are executed, C++ evaluates only as much of the expression as necessary. For example, if A is false, then the expression A && B has to be false regardless of the value of B, and there is no need to evaluate B. Similarly, if A is true, then the expression A // B is true, and there is no need to evaluate B.

Precedence and Associativity

A condition can contain several logical operators as in the following:

```
!(b==c || b==5.5)
```

The hierarchy, from highest to lowest, is !, &&, //, but parentheses can be used to change the hierarchy. In this example, the expressions b==c and b==5.5 are evaluated first. Suppose b is equal to 3 and c is equal to 5. Then neither expression is true, so the expression b==c // b==5.5 is false. Then the ! operator is applied to the false condition, which gives a true condition.

Blanks cannot be used to separate the characters in either // or &&. A common error is to use = instead of == in a logical expression.

TABLE 3-1   Logical Operators

A	B	A && B	A // B	!A	!B
False	False	False	False	True	True
False	True	False	True	True	False
True	False	False	True	False	True
True	True	True	True	False	False

TABLE 3-2   Operator Precedence for Arithmetic,
Relational, and Logical Operators

PRECEDENCE	OPERATION	ASSOCIATIVITY
1	( )	innermost first
2	`+` `-` `++` `--` `(type)` `!`	right to left (unary)
3	`*` `/` `%`	left to right
4	`+` `-`	left to right
5	`<` `<=` `>` `>=`	left to right
6	`==` `!=`	left to right
7	`&&`	left to right
8	`//`	left to right
9	`=` `+=` `-=` `*=` `/=` `%=`	right to left

A condition can contain both arithmetic operators and relational operators, as well as logical operators. Table 3-2 gives the precedence and the associativity order for the elements in a condition.

PRACTICE!

Determine if the following conditions are true or false. Assume that the following variables have been declared and given these values:
a $\boxed{5.5}$ b $\boxed{1.5}$ k $\boxed{3}$

1. `a < 10.0+k`

2. `a+b >= 6.5`

3. `!(a == 3*b)`

4. `-k <= k+6`

5. `a<10 && a>5`

6. `fabs(k)>3 // k<b-a`

3.3 SELECTION STATEMENTS

The `if` statement tests a condition and then perform statements based on whether the condition is true or false. C++ contains two forms of if statements—the simple `if` statement and the `if/else` statement.

Simple `if` Statement

The simplest form of an `if` statement has the following general form:

```
if (condition)
    statement 1;
```

If the condition is true, the program executes statement 1. If the condition is false, the program skips statement 1.

The statement within the if statement is indented so that it is easier to visualize the structure of the program from the statements.

To execute several statements (or a sequence structure) if the condition is true, use a **compound statement**, or **block**, which is composed of a set of statements enclosed in braces. The location of the braces is a matter of style. Here are two common styles:

STYLE 1	STYLE 2
`if (condition)`	`if (condition) {`
`{`	`statement 1;`
`statement 1;`	`statement 2;`
`statement 2;`	`. . .`
`. . .`	`statement n;`
`statement n;`	`}`
`}`	

Solutions in this text, follow the first style convention, with both braces on lines by themselves. Although this makes the program a little longer, it also makes it easier to notice if you omit a brace by accident. Figure 3.5 shows flowcharts of the control flow with simple if statements containing either one statement to execute or several statements to execute if the condition is true.

Here is a specific example of an `if` statement follows:

```
if (a < 50)
{
    ++count;
    sum += a;
}
```

If `a` is less than 50, then `count` is incremented by 1, and a is added to `sum`; otherwise, these two statements are skipped.

Figure 3.5. Flowcharts for selection statements.

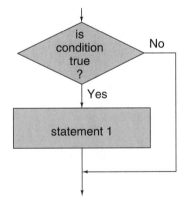

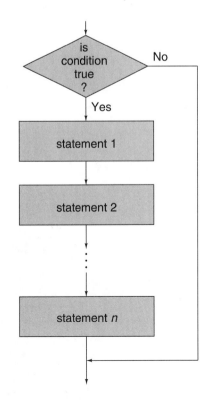

Note that `if` statements can also be nested. The following example includes an `if` statement within an `if` statement:

```
if (a < 50)
{
    ++count;
    sum += a;
    if (b > a)
        b = 0;
}
```

If `a` is less than 50, the program increments count by 1 and adds `a` to `sum`. In addition, if `b` is greater than `a`, the program sets `b` to zero. If `a` is not less than 50, then all of these statements are skipped. Be sure to indent the statements in each if statement when they are nested.

`if/else` Statement

An `if/else` statement executes one set of statements if a condition is true, and a different set of statements if the condition is false. The simplest form of an `if/else` statement is the following:

```
if (condition)
    statement 1;
else
    statement 2;
```

Statements 1 and 2 can be replaced by compound statements. Statement 1 or statement 2 can also be an **empty statement**, which is a semicolon. If statement 2 is an empty statement, the `if/else` statement should probably be posed as a simple if statement. There are situations in which it is convenient to use an empty statement for statement 1; however, these statements can also be rewritten as a simple if statement with the condition reversed. For example, the following two statements are equivalent:

```
if (a < b)                  if (a >= b)
    ;                           count++;
else
    count++;
```

Consider this `if/else` statement:

```
if (d <= 30)
    velocity = 0.425 + 0.00175*d*d;
else
    velocity = 0.625 + 0.12*d - 0.0025*d*d;
```

In this example, `velocity` is computed with the first assignment statement if the distance d is less than or equal to 30; otherwise, `velocity` is computed with the second assignment statement. A flowchart for this `if/else` statement is shown in Figure 3.6.

Here is another example of the `if/else` statement:

```
if (fabs(denominator) < 0.0001)
    cout << "Denominator close to zero";
else
{
    fraction = numerator/denominator;
    cout << "fraction = " << fraction << endl;
}
```

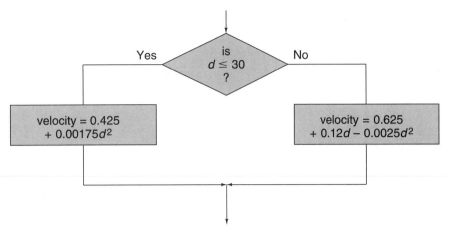

Figure 3.6. Flowchart for selection structure.

In this example, the program examines the absolute value of the variable `denominator`. If this value is close to zero, a message is printed indicating that the division cannot be performed. If the value of `denominator` is not close to zero, the program computes and prints the value of `x`. The flowchart for this statement was shown in Figure 3.3.

Consider the following set of nested `if/else` statements:

```
if (x > y)
    if (y < z)
        k++;
      else
          m++;
    else
       j++;
```

The value of `k` is incremented when `x > y` and `y < z`. The value of `m` is incremented when `x > y` and `y >= z`. The value of `j` is incremented when `x <= y`. With careful indenting, this statement is straightforward to follow. Suppose that you now eliminate the `else` portion of the inner `if` statement. If you keep the same indention, the statements become the following:

```
if (x > y)
    if (y < z)
        k++;
    else
       j++;
```

It might appear that `j` is incremented when `x <= y`, but that is not correct. The C++ compiler will associate an `else` statement with the closest `if` statement within a block. Therefore, no matter what indenting is used, the statement above is executed as if it were the following:

```
if (x > y)
    if (y < z)
        k++;
    else
       j++;
```

Thus, j is incremented when x > y and y > = z. If you intend for j to be incremented when x < = y, then you must use braces to define the inner statement as a block:

```
if (x > y)
{
   if (y < z)
      k++;
}
else
   j++;
```

To avoid confusion and possible errors when using nested if/else statements, you should routinely use braces to clearly define the blocks of statements that go together.

This section has shown, a number of ways to compare values in selection statements. A caution is necessary when comparing floating-point values. For example, in an example in this section, we did not compare denominator to zero, but instead used a condition to see if the absolute value of denominator were less than a small value. Similarly, if you wanted to know if y was close to the value 10.5, you should use a condition such as fabs(y-10.5) < = 0.0001 instead of y == 10.5. In general, do not use the equality operator with floating-point values.

PRACTICE!

In problems 1 through 6, give the corresponding C++ statements. Assume that the variables have been declared and have reasonable values.

1. If time is greater than 15.0, increment time by 1.0.
2. When the square root of poly is less than 0.5, print the value of poly.
3. If the difference between volt_1 and volt_2 is larger than 10.0, print the values of volt_1 and volt_2.
4. If the natural logarithm of x is greater than or equal to 3, set time equal to zero and decrement count.
5. If dist is less than 50.0 and time is greater than 10.0, increment time by 2; otherwise, increment time by 2.5.
6. If dist is greater than or equal to 100.0, increment time by 2.0. If dist is between 50 and 100, increment time by 1. Otherwise, increment time by 0.5.

3.4 LOOP STRUCTURES

Loops are used for the repetition control structures. C++ contains three different types of loops—the while loop, do/while loop, and for loop. In addition, two additional statements can be used with loops to modify their performance—the break statement and the continue statement.

Before examining these loop structures, here are two debugging suggestions that are useful when you look for errors in programs that contain loops. When compiling longer programs, it is not uncommon to have a large number of compiler errors. Rather than trying to find each error separately, it is better to recompile your program after correcting several obvious syntax errors. One error will often generate several error messages. Some of these error messages may describe errors that aren't in your program at all, but were printed because the original error confused the compiler.

The second debugging suggestion relates to errors inside a loop. When you want to determine if the steps in a loop are working the way that you want, include cout statements in the loop to provide a memory snapshot of key variables each time the loop is executed. Then, if there is an error, you have much of the information that you need to determine what is causing the error.

while Loop

The general form of a while loop is:

```
while (condition)
    statement;
```

The condition is evaluated before the statement within the loop is executed. (The statement may also be a compound statement.) If the condition is false, the loop statement is skipped, and execution continues with the statement following the while loop. If the condition is true, the loop statement is executed, and the condition is evaluated again. If it is still true, then the statement is executed again, and the condition is evaluated again. This repetition continues until the condition is false.

The statement within the loop must modify variables that are used in the condition. Otherwise, the value of the condition will never change, and the program will either never execute the statements in the loop or will never be able to exit the loop. An **infinite loop** is generated if the condition in a while loop is always true. Most computers have a system-defined limit on the amount of time that can be used by a program, and will generate an execution error when this limit is exceeded. Other systems require that the user enter a special set of characters, such as the control key followed by the character c (abbreviated as ^c), to stop or abort the execution of a program. Nearly everyone eventually writes a program that inadvertently contains an infinite loop, so be sure you know the special characters to stop the program for your system.

The following pseudocode and program use a while loop to generate a conversion table for converting degrees to radians. The degree values start at 0°, increment by 10°, and go through 360°.

Refinement in Pseudocode:

```
main:  set degrees to zero
       while degrees # 360
           convert degrees to radians
           print degrees, radians
           add 10 to degrees
```

```
//----------------------------------------------------------
//  Program chapter3_1
//  /
//  This program prints a degree-to-radian table
//  using a while loop structure.

#include <iostream.h>
#include <iomanip.h>
#include <stdlib.h>

int main()
{
    //  Define constant and variables.
    const double PI=3.141593;
    int degrees=0;
    double radians;

    //  Print degrees and radians in a loop.
    cout << "Degrees to Radians" << endl;
    while (degrees <= 360)
```

```
    {
        radians = degrees*PI/180;
       cout << degrees << "   "
              << setprecision(4) << radians << endl;
       degrees += 10;
    }

    //  Exit program.
    return EXIT_SUCCESS;
}
//-----------------------------------------------------------
```

The first few lines of output from the program follow:

```
Degrees to Radians

0   0
10   0.1745
20   0.3491
 .
 .
 .
```

do/while Loop

The do/while loop is similar to the while loop, except that the condition is tested at the end of the loop instead of at the beginning. Because the condition is tested at the end of the loop, the do/while loop is always executed at least once. A while loop, on the other hand, may not be executed at all if the condition is initially false. The general form of the do/while loop is

```
do
    statement;
while  (condition);
```

The following pseudocode and program print the degree-to conversion table using a do/while loop instead of a while loop:

Refinement in Pseudocode:

main: set degrees to zero
do
    convert degrees to radians
    print degrees, radians
    add 10 to degrees
while degrees ≤ 360

```
//-----------------------------------------------------------
//  Program chapter3_2
/ /
//  This program prints a degree-to-radian table
//  using a do-while loop structure.

#include <iostream.h>
#include <iomanip.h>
#include <stdlib.h>

int main()
{
    //  Define constant and variables.
    const double PI=3.141593;
    int degrees=0;
    double radians;
```

```
    //  Print degrees and radians in a loop.
    cout << "Degrees to Radians" << endl;
  do
  {
      radians = degrees*PI/180;
      cout << degrees << "   "
           << setprecision(4) << radians << endl;
      degrees += 10;
  } while (degrees <= 360);
  //  Exit program.
  return EXIT_SUCCESS;
}
//-----------------------------------------------------------
```

for Loop

Many programs require loops that are based on the value of a variable that increments (or decrements) by the same amount each time through the for loop. When the variable reaches a specified value, the program should exit the loop. This type of loop can be implemented as a while loop, but it can also be easily implemented with the for loop. The general form of the for loop is

```
for (expression_1; expression_2; expression_3)
    statement;
```

Expression_1 is used to initialize the **loop-control variable**. Expression_2 specifies the condition that should be true to continue the loop repetition. Expression_3 specifies the modification to the loop-control variable.

For example, if you want to execute a loop 10 times, with the value of the variable k going from 1 to 10 in increments of 1, you could use the following:

```
for (int k=1; k<11; k++)
    statement;
for (int k=1; k<=10; k++)
    statement;
```

To execute a loop with the value of the variable n going from 20 to 0 in increments of −2, you could use this loop structure:

```
for (int n=20; n>=0; n=n-2)
    statement;
```

The for loop could also have been written in this form:

```
for (int n=20; n>=0; n-=2)
    statement;
```

Both forms are valid, but the abbreviated form is commonly used because it is shorter.

The following expression computes the number of times that a for loop will be executed:

$$(\text{floor})\left(\frac{\text{final value} - \text{initial value}}{\text{increment}}\right) + 1$$

If this value is negative, the loop is not executed. Thus, if a for statement has the following structure:

```
for (int k=5; k<=83; k+=4)
    statement;
```

it would be executed the following number of times:

$$\text{floor}\left(\frac{83-5}{4}\right) + 1 \ = \ \text{floor}\left(\frac{78}{4}\right) + 1 \ = \ 20$$

The value of k would be 5, and then 9, and then 13, and so on, until the final value of 81. The loop would not be executed with the value of 85 because the loop condition is not true when k is equal to 85.

The following pseudocode and program print the degree-to-radian conversion table shown earlier with a while loop, now modified to use a for loop. Note that the pseudocode for the while loop solution to this problem and the pseudocode for the do loop solution to this problem are identical.

Refinement in Pseudocode:

main: set degrees to zero
while degrees ≤ 360
   convert degrees to radians
   print degrees, radians
   add 10 to degrees

```
//-----------------------------------------------------------
//  Program chapter3_3
// /
//  This program prints a degree-to-radian table
// using a for loop structure.

#include <iostream.h>
#include <iomanip.h>
#include <stdlib.h>

int main()
{
   //  Define constant and variable.
  const double PI=3.141593;
   double radians;

   //  Print degrees and radians in a loop.
  cout << "Degrees to Radians" << endl;
   for (int degrees=0; degrees<=360; degrees+=10)
 {
     radians = degrees*PI/180;
    cout << degrees << "   "
          << setprecision(4) << radians << endl;
 }

   //  Exit program.
   return EXIT_SUCCESS;
}
//-----------------------------------------------------------
```

Note that the value of degrees did not need to be initialized in the declaration because it is initialized in the for loop statement.

PRACTICE!

Determine the number of times that the for loops below are executed.

1.
```
for (k=3; k<=20; k++)
   statement;
```
2.
```
for (k=3; k<=20; ++k)
   statement;
```
3.
```
for (count=-2; count<=14; count++)
   statement;
```
4.
```
for (k=-2; k>=-10; k--)
   statement;
```
5.
```
for (time=10; time>=0; time--)
   statement;
```
6.
```
for (time=10; time>=5; time++)
   statement;
```

break and continue Statements

The break statement can be used with any of the loop structures presented. It is used to exit from a loop. In contrast, the continue statement is used to skip the remaining statements in the current pass or **iteration** of the loop, and then continue with the next iteration of the loop. Thus, in a while or do/while loop, the condition is evaluated after the continue statement is executed to determine if the statements in the loop are to be executed again. In a for loop, the loop-control variable is modified, and the condition is evaluated to determine if the statements in the loop are to be executed again. Both the break and continue statements are useful in exiting either the current iteration or the entire loop when error conditions are encountered.

To illustrate the difference between the break and the continue statements, consider the following loop that reads values from the keyboard:

```
sum = 0;
for (int k=1; k<21; k++)
{
   cin >> x;
   if (x > 10)
      break;
   sum += x;
}
cout << "sum = " << sum << endl;
```

This loop reads up to 20 values from the keyboard. If all 20 values are less than or equal to 10.0, the statements compute the sum of the values and print the sum. But, if a value is read that is greater than 10.0, the break statement causes control to break out of the loop and execute the cout statement. Thus, the sum printed is only the sum of the values up to the value greater than 10.0.

Now, consider this variation of the previous loop:

```
sum = 0;
for (int k=1; k<21; k++)
{
```

```
      cin >> x;
       if (x > 10)
            continue;
        sum += x;
    }
    cout << "sum =  " << sum << endl;
```

In this loop, the sum of all 20 values is printed if all values are less than or equal to 10.0. However, if a value is greater than 10.0, the `continue` statement causes control to skip the rest of the statements in that iteration of the loop, and to continue with the next iteration. So the sum printed is the sum of all values in the 20 values that are less than or equal to 10.

3.5 PROBLEM SOLVING APPLIED: WEATHER BALLOONS

Weather balloons are used to gather temperature and pressure data at various altitudes in the atmosphere. The balloon rises because the density of the helium in the balloon is less than the density of the surrounding air outside. As the balloon rises, the surrounding air becomes less dense; thus, the balloon's ascent slows until it reaches a point of equilibrium. During the day, sunlight warms the helium trapped inside the balloon, which causes the helium to expand and become less dense, and the balloon to rise higher. During the night, however, the helium in the balloon cools and becomes more dense, causing the balloon to descend to a lower altitude. The next day, the sun heats the helium again and the balloon rises. Over time this process generates a set of altitude measurements that can be approximated with a polynomial equation.

Assume that the following polynomial represents the altitude or height in meters during the first 48 hours following the launch of a weather balloon:

$$\text{alt}(t) = -0.12t + 12t^3 - 380t^2 + 4100t + 220$$

where the units of t are hours. The corresponding polynomial model for the velocity in meters per hour of the weather balloon is as follows:

$$v(t) = -0.48t^3 + 36t^2 - 760t + 4100$$

Print a table of the altitude and the velocity for this weather balloon using units of meters and meters/second. Let the user enter the start time, increment in time between lines of the table, and ending time, where all the time values must be less than 48 hours. In addition to printing the table, also print the peak altitude from the table and its corresponding time.

1. Problem Statement

Using the polynomials that represent the altitude and velocity for a weather balloon, print a table using units of meters and meters/second. Also find the maximum altitude (or height) and its corresponding time.

2. Input/Output Description

The following I/O diagram shows the user input that represents the starting time, time increment, and ending time for the table. The output is the table of altitude and velocity values and the maximum altitude and its corresponding time.

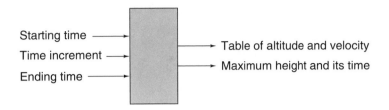

3. Hand Example

Assume that the starting time is 0 hours, the time increment is 1 hour, and the ending time is 5 hours. To obtain the correct units, you need to divide the velocity value in meters/hour by 3600 in order to get meters/sec. Using a calculator, you can then compute the following values:

Time	Altitude (m)	Velocity(m/s)
0	220.00	1.14
1	3,951.88	0.94
2	6,994.08	0.76
3	9,414.28	0.59
4	11,277.28	0.45
5	12,645.00	0.32

You can also determine the maximum altitude from this table, which is 12,645.00 meters; it occurred at 5 hours.

4. Algorithm Development

First develop the decomposition outline, because it breaks the solution into a series of sequential steps.

Decomposition Outline

1. Get user input to specify times for the table.
2. Generate and print conversion table, and find maximum height and corresponding time.
3. Print maximum height and corresponding time.

The second step in the decomposition outline represents a loop, in which the program generates the table and, at the same time, keeps track of the maximum height. As you refine this outline, and particularly step 2, into more detail, think carefully about finding the maximum height. Look back at the hand example. Once the table has been printed, it is easy to look at it and select the maximum height. However, when the computer is computing and printing the table, it does not have all the data at one time, but only the information for the current line in the table. Therefore, to keep track of the maximum, you need to specify a separate variable to store the maximum value. Each time that you compute a new height, you will compare that value to the maximum value. If the new value is larger, replace the maximum with this new value. You also need to keep track of the corresponding time. The following refinement in pseudocode outlines these new steps:

Refinement in Pseudocode:

main: read initial, increment, final values from keyboard
   set max_height to zero
   set max_time to zero
   print table heading
   set time to initial
   while time <= final
       compute height and velocity
       print height and velocity
       if height > max_height
           set max_height to height
           set max_time to time
       add increment to time
   print max_time and max_height

The steps in the pseudocode are now detailed enough to convert into C++. Note that velocity is converted from meters/hour to meters/second in the `cout` statement.

```cpp
//--------------------------------------------------------
//  Program chapter3_4
/ /
//  This program prints a table of height and
//  velocity values for a weather balloon.

#include <iostream.h>
#include <iomanip.h>
#include <stdlib.h>
#include <math.h>

int main()
{
  //  Define variables.
  double initial, increment, final;

  //  Get user input.
  cout << "Enter initial value for table (in hours): ";
  cin >> initial;
  cout << "Enter increment between lines (in hours): ";
  cin >> increment;
  cout << "Enter final value for table (in hours): ";
  cin >> final;

  //  Print report heading.
  cout << endl
       << "Weather Balloon Information" << endl;
  cout << "Time     Height    Velocity" << endl;
  cout << "(hr)     (m)        (m/s)" << endl;

  //  Compute and print report.
  double height, velocity, max_time=0, max_height=0;
  for (double time=initial; time<=final; time+=increment)
  {
     height = -0.12*pow(time,4) + 12*pow(time,3)
           - 380*time*time + 4100*time + 220;
     velocity = -0.48*pow(time,3) + 36*time*time
           - 760*time + 4100;
    cout << setw(4) << setprecision(2) << time << "    "
         << setw(8) << setprecision(2) << height << "    "
          << setprecision(2) << velocity/3600 << endl;
```

```
   if (height > max_height)
 {
     max_height = height;
   max_time = time;
  }
 }
  //  Print maximum height and corresponding time.
 cout << endl << "Maximum balloon height was "
      << setprecision(2) << max_height
      << " m" << endl;
 cout << "and it occurred at " << setprecision(2)
      << max_time << " hr." << endl;

 // Exit program.
 return EXIT_SUCCESS;
}
//----------------------------------------------------
```

5. Testing
Using data from the hand example, you will have the following interaction with the program:

```
Enter initial value for table (in hours): 0
Enter increment between lines (in hours): 1
Enter final value for table (in hours): 5

Weather  Balloon  Information
Time     Height   Velocity
(hr)     (m)      (m/s)
   0       220    1.14
   1    3.95e+03   0.94
   2    6.99e+03   0.76
   3    9.41e+03   0.59
   4    1.13e+04   0.45
   5    1.26e+04   0.32

Maximum balloon height was 1.26e+04 m
and it occurred at 5 hr.
```

Figure 3.7 contains a plot of the altitude and velocity of the balloon for a period of 48 hours. From the plots, you can see the periods during which the balloon rises or falls.

3.6 DATA FILES

The solutions to engineering problems often involve a lot of data. These data can be generated by the program as output, or can be input data used by the program. It is not generally feasible either to print large amounts of data to the screen, or to read large amounts of data from the keyboard. In these cases, **data files** are generally used to store the data.

Data files are similar to the program files that you create to store C++ programs. In fact, a C++ program file is an input data file to the C++ compiler, and the object program is an output file from the C++ compiler. This section takes a look at the C++ statements for interacting with data files and gives examples that generate and read information from data files.

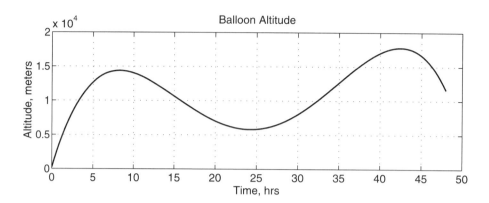

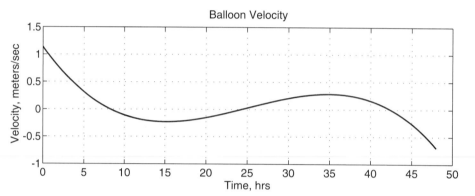

Figure 3.7. Weather balloon altitude and velocity.

When debugging programs that read information from data files, echo (or print) the information read from the file to be sure that the data are being read properly. If the data values are all zero, or are unusual numbers, it may be possible that the program can't find the file because it's in a directory that the program cannot access. The solution is either to move the file to a directory that the program can access, or to change some of the operating system parameters so that the program can find the file.

The examples that follow use data files containing sensor data from a **seismometer**. Seismometers are usually buried near the surface of the earth and record earth motion. They are very sensitive, and can record tidal motion even though they may be located hundreds of miles from the ocean. Seismometer data are collected from sensors all over the earth, and are sent by satellite to central locations for collection and analysis. By studying this motion, scientists and engineers may be able to predict earthquakes from seismometer data.

I/O Statements

In order to access a data file, we must include the preprocessor directive

```
#include <fstream.h>
```

where the header file `fstream.h` contains information related to manipulating a file.

Each data file used in a program needs a **file object** to attach the data file to the program. If a program uses two files, then each should be given a different file object

name for readability. We use the following statement to create a file object named
`sensor1`:

```
fstream sensor1;
```

After a file object is defined, it can be associated with a specific file using the `open`
function. The two arguments for this function are the file name, which needs to be enclosed
in double quotes, and the file status (also called the **file access mode**). The access mode
tells the system whether the file is an input file or an output file. If we are going to read information from a file with a program, the file access mode is `ios::in`. If we are going to write
information to a file with a program, the file access mode is `ios::out`. Thus, the following
statement specifies that the file object `sensor1` is going to be used with a file named
`sensor1.dat`, from which we will read information:

```
sensor1.open("sensor1.dat", ios::in);
```

The general form of the statements used to create a file object and attach it to a
data file are as follows:

```
fstream object_name;
object_name.open("file_name", access mode);
```

These two statements can be combined in the statement

```
fstream object_name("file_name", access mode);
```

Since some operating systems are not case sensitive, we will use all lowercase letters in
file names to avoid any potential problems.

Once an input file and its object have been specified, we can read information
from the file much as we would read information from the keyboard. However, instead
of using `cin`, we use the file object; the `>>` operator is still cascaded to receive data
from the file. If each line in the `sensor1.dat` file contains a time and a sensor reading, we can read one line of this information and store the values in the variables `time`
and `motion` with the file input statement

```
sensor1 >> time >> motion;
```

The difference between the file input statement and the `cin` statement is that the first
word in the file input statement is the file object name; otherwise, both statements are the
same. The `cin` statement converts the characters received from the keyboard to values,
and the file input statement converts the characters from the lines in the data file to values.

In a similar manner, if the file is an output file, we can write information to it
with the file output statement. The first word of the file output statement is also the
file object name, and the rest of the statement is the same as the `cout` statement. The
`<<` operator is still used to send data to the file. For example, consider Program
`chapters3_4`, which computed and printed a table of time, altitude, and velocity
data. If we wanted to modify this program so that it generated a data file containing the
same data, we could use an object name `balloon` that would be associated with an
output file named `balloon.dat` using the statements

```
fstream balloon;
balloon.open("balloon.dat", ios::out);
```

where the file access mode is `ios::out`. Then, as we compute the information on
time, height, and velocity, we can write it to the file with the file output statement

```
balloon << time << height << velocity/3600 << endl;
```

The `endl` reference causes a skip to a new line after each group of three values is written to the file.

The `close` function, used to close a file after we are finished with it, does not have an argument. Its general form is

```
object_name.close()
```

Thus, to close the two files used in the examples we have been considering, we can use the following statements:

```
sensor1.close();
balloon.close();
```

There is no distinction between closing an input file and closing an output file. If a file has not been closed when the `return` statement is executed, it will automatically be closed. It is important to note that executing the `close` function does not destroy our file object; it only closes the data file. The object still exists, and therefore, we could use another `open` function to attach it to a new data file.

Reading Data Files

In order to read information from a data file, you first need to know some details about the file. Obviously you have to know the file name, to use the `fopen` statement to associate the file with its pointer. You also need to know the order and data type of the values stored in the file, in order to declare corresponding identifiers correctly.

In addition, you need to know if there is any special information in the file to help determine just how much information the file contains. If you attempt to execute an `fscanf` statement after reading all the data in the file, an error occurs. In order to avoid this error, you need to know when we have read all the data.

Data files generally have one of three common structures. Some files have been generated so that the first line in the file contains the number of lines (also called **records**) with information that follow. For example, suppose that a file containing sensor data has 150 sets of time and sensor information. The data file could be constructed so that the first line contains only the value 150; that line would then be followed by 150 lines containing the sensor data. To read the data from this file, you would read the value from the first line in the file and then use a `for` loop to read the rest of the information. This type of loop is also called a **counter-controlled loop.**

Another form of file structure uses a **trailer** or **sentinel signal.** These signals are special data values used to indicate or signal the last record of a file. For example, the sensor data file constructed with a sentinel signal would contain the 150 lines of information followed by a line with special values, such as −999.0 for the time and sensor value. These sentinel signals must be values that could not appear as regular data, in order to avoid confusion. To read data from this type of file, use a `while` loop with a condition that is true as long as the data value is not the sentinel signal. This type of loop is called a **sentinel-controlled loop.**

The third data file structure contains only valid data. It does not contain an initial line with the number of valid data records that follow, and it does not contain a trailer or sentinel signal. For this type of data file, use the value returned by the `fscanf` function to help determine when you have read the last line of the file. To read data from this type of file, use a `while` loop with a condition that is true as long as you are not at the end of the file.

Here are programs for reading sensor information and printing a summary report that contains the number of sensor readings, average value, maximum value, and minimum value. Each of the three common file formats discussed will be used in the programs.

Specified Number of Records   Assume that the first record in the data file `sensor1.dat` contains an integer that specifies the number of records of sensor information that follow. Each following line contains a time and sensor reading, as shown below:

```
10
0.0        132.5
0.1        147.2
0.2        148.3
0.3        157.3
0.4        163.2
0.5        158.2
0.6        169.3
0.7        148.2
0.8        137.6
0.9        135.9
```

The process of first reading the number of data points, and then using that to specify the number of times to read data and accumulate information, is easily described using a variable-controlled loop. In the following program, the first actual data value is used to initialize the `maximum` and `minimum` values. If you set the minimum value initially to zero and all the sensor values were greater than zero, the program would print the erroneous value of zero for the `minimum` sensor reading.

```
//-------------------------------------------------------------
//  Program chapter3_5
//  /
//  This program generates a summary report from
//  a data file that has the number of data points
//  in the first record.

#include <iostream.h>
#include <fstream.h>
#include <stdlib.h>

int main()
{
    //  Open file and read the number of data points.
    int num_data_pts;
    fstream sensor1;
    sensor1.open("sensor1.dat", ios::in);
    sensor1 >> num_data_pts;
```

```
    //  Read data and compute summary information.
    double time, motion, max, min, sum;
    sensor1 >> time >> motion;
    max = min = sum = motion;
    for (int k=2; k<=num_data_pts; k++)
    {
        sensor1 >> time >> motion;
        sum += motion;
        if (motion > max)
           max = motion;
        if (motion < min)
           min = motion;
    }

    //  Print summary information.
    cout << "Number of readings:  "
         << num_data_pts << endl;
    cout << "Average reading:  "
         << sum/num_data_pts << endl;
    cout << "Maximum reading:  " << max << endl;
    cout << "Minimum reading:  " << min << endl;

    //  Close file and exit program.
    sensor1.close();
    return EXIT_SUCCESS;
}
//------------------------------------------------------------
```

The report printed by this program is the following:

```
Number of readings:    10
Average reading:    149.77
Maximum reading:     169.3
Minimum reading:     132.5
```

Trailer or Sentinel Signals   Assume that the data file `sensor2.dat` contains the same information as the `sensor1.dat` file, but instead of giving the number of valid data records at the beginning of the file, a final record contains a trailer signal. The time value on last line in the file will contain a negative value, so we know that it is not a valid line of information. A second number must be included on the trailer line because the statement that reads each line expects two values; otherwise an error occurs. The contents of the data file are as follows:

```
0.0        132.5
0.1        147.2
0.2        148.3
0.3        157.3
0.4        163.2
0.5        158.2
0.6        169.3
0.7        148.2
0.8        137.6
0.9        135.9
-99        -99
```

The process of reading and accumulating information until the program reads the trailer signal is easily described using a do/while loop structure, as shown in the following program.

```
//------------------------------------------------------------
//  Program chapter3_6
//
//  This program generates a summary report from
// a data file that has a trailer record with
//  negative values.

#include <iostream.h>
#include <fstream.h>
#include <stdlib.h>

int main()
{
   //  Open file and read the first set of data.
   double time, motion;
   fstream sensor2;
    sensor2.open("sensor2.dat", ios::in);
   sensor2 >> time >> motion;

   //  Define and initialize variables.
   int num_data_pts=0;
   double max=motion, min=motion, sum=0;

   //  Update summary data until trailer record read.
   while (time >= 0)
   {
      sum += motion;
      if (motion > max)
         max = motion;
      if (motion < min)
         min = motion;
      num_data_pts++;
      sensor2 >> time >> motion;
   }

   //  Print summary information.
   cout << "Number of readings:  "
        << num_data_pts << endl;
   cout << "Average reading:  "
        << sum/num_data_pts << endl;
   cout << "Maximum reading:  " << max << endl;
   cout << "Minimum reading:  " << min << endl;
   //  Close file and exit program.
   sensor2.close();
   return EXIT_SUCCESS;
}
//------------------------------------------------------------
```

The report printed by this program using the sensor2.dat file is exactly the same as the report printed using the sensor1.dat file.

End-of-File   A special end-of-file indicator is inserted at the end of every data file. The eof function (accessed through the iostream header file) can be used to detect when this indicator has been reached in reading a data file. Thus, if we want to read information until we run out of data, we can use the `eof` function in the condition for a `while` loop to allow us to continue reading and processing information as long as we are not at the end of the file.

Let us assume that the data file `sensor3.dat` contains the same information as the file `sensor2.dat`, except that it does not include the trailer signal. Then the contents of `sensor3.dat` are as follows:

```
0.0        132.5
0.1        147.2
0.2        148.3
0.3        157.3
0.4        163.2
0.5        158.2
0.6        169.3
0.7        148.2
0.8        137.6
0.9        135.9
```

The following program reads and accumulates information until it reaches the end of the data file.

```cpp
//------------------------------------------------------------
//   Program chapter3_7
/ /
//   This program generates a summary report from a data
//   file that contains only data.  The file does not
//    contain initial information or trailer information.

#include <iostream.h>
#include <fstream.h>
#include <stdlib.h>

int main()
{
   //   Open file and read the first set of data.
   double time, motion;
   fstream sensor3;
     sensor3.open("sensor3.dat", ios::in);
   sensor3 >> time >> motion;

   //   Define and initialize variables.
   int num_data_pts=0;
   double max=motion, min=motion, sum=0;

   //   While not at the end of the file,
    //   read and accumulate information.
   while (!sensor3.eof())
   {
      sum += motion;
      if (motion > max)
         max = motion;
```

```
          if (motion < min)
             min = motion;
           num_data_pts++;
          sensor3 >> time >> motion;
   }

      //  Print summary information.
      cout << "Number of readings:  "
           << num_data_pts << endl;
      cout << "Average reading:  "
            << sum/num_data_pts << endl;
      cout << "Maximum reading:  " << max << endl;
      cout << "Minimum reading:  " << min << endl;

      //  Close file and exit program.
       sensor3.close();
     return EXIT_SUCCESS;
   }
   //-----------------------------------------------------------
```

The programs in this section work properly if the data files exist and contain the expected information. A division-by-zero error will occur if the number of points is zero. If the program attempts to open a file that does not exist, or that cannot be found because the system is looking on the wrong disk drive, an error will also occur. The `fail` **function** (also accessed through the `iostream` header file) can be used to determine the success or failure of an open operation; the function returns a true value if the operation failed and a false value if it was successful. In the former case, an appropriate error message can be printed before ending the program

Generating a Data File

Generating a data file is very similar to printing a report. Instead of writing the line to the terminal screen, the program writes it to a data file. Before generating the data file, though, you must decide what file structure you want to use. The previous discussion, presented the three most common file structures—files with an initial record giving the number of valid records that follow, files with a trailer or sentinel record to indicate the end of the valid data, and files with only valid data records and no special beginning or ending records.

There are advantages and disadvantages to each of the three file structures. A file with a trailer signal is simple to use, but choosing a value for the trailer signal must be done carefully so that it does not contain values that could occur in the valid data. If the first record in the data file will contain the number of lines of actual data, you must know how many lines of data will be in the file before you begin to generate the file. It may not always be easy to determine the number of lines before executing the program that generates the file. The simplest file to generate is the one that contains only the valid information, with no special information at the beginning or end of the file. If the information in the file is going to be used with a plotting package, it is usually best to use this third file structure, which includes only valid information.

Here is a modification of the program presented earlier in the chapter, which printed a table of time, altitude, and velocity values for a weather balloon. In addition to generating a table of information displayed on the screen, the modified program writes the information to a data file. Compare this program to the one in Section 3.5.

```
//-------------------------------------------------------
//   Program chapter3_8
//  /
//   This program generates a file of height and
//   velocity values for a weather balloon. The
//   information is also printed in a report.

#include <iostream.h>
#include <iomanip.h>
#include <fstream.h>
#include <stdlib.h>
#include <math.h>

int main()
{
   //  Define variables.
   double initial, increment, final;

   //  Get user input.
   cout << "Enter initial value for table (in hours): ";
 cin >> initial;
   cout << "Enter increment between lines (in hours): ";
  cin >> increment;
   cout << "Enter final value for table (in hours): ";
 cin >> final;

   //  Print report heading.
   cout << endl
         << "Weather Balloon Information" << endl;
 cout << "Time      Height     Velocity" << endl;
 cout << "(hr)      (m)        (m/s)" << endl;

   //  Open output file.
   fstream balloon;
    balloon.open("balloon.dat", ios::out);

   //  Compute and print report
  //   and write data to a file.
   double height, velocity, max_time=0, max_height=0;
   for (double time=initial; time<=final; time+=increment)
 {
      height = -0.12*pow(time,4) + 12*pow(time,3)
            - 380*time*time + 4100*time + 220;
      velocity = -0.48*pow(time,3) + 36*time*time
            - 760*time + 4100;
    cout << setw(4) << setprecision(2) << time << "    "
         << setw(8) << setprecision(2) << height << "    "
          << setprecision(2) << velocity/3600 << endl;
    balloon << time << "   " << height << "   "
           << velocity/3600 << endl;
     if (height > max_height)
   {
       max_height = height;
        max_time = time;
   }
 }

   //  Print maximum height and corresponding time.
  cout << endl << "Maximum balloon height was "
       << setprecision(2) << max_height
```

```
          << " m" << endl;
     cout << "and it occurred at " << setprecision(2)
          << max_time << " hr." << endl;

     //  Close file and exit program.
     balloon.close();
     return EXIT_SUCCESS;
}
//-----------------------------------------------------------
```

The first few lines of a data file generated by this program using an initial time of 0 hours, an increment of 0.5 hours, and a final time of 48 hours are

```
0.00                   220.00              1.14
0.50                   2176.49             1.04
1.00                   3951.88             0.94
1.50                   5554.89             0.84
...
```

A plot of this specific file was shown in Figure 3.7.

C++ STATEMENT SUMMARY

Declaration for file object

```
     fstream sensor1;
```

if statement:

```
     if (temp > 100)
         cout << "Temperature exceeds limit" << endl;
```

if/else statement:

```
     if (d <= 30)
         velocity = 4.25 + 0.00175*d*d;
     else
         velocity = 0.65 + 0.12*d - 0.0025*d*d;
```

while loop:

```
     while (degrees <= 360)
     {
         radians = degrees*PI/180;
         cout << degrees << " " << radians << endl;
         degrees += 10;
     }
```

do/while loop:

```
     do
     {
          radians = degrees*PI/180;
          cout << degrees << " " << radians << endl;
          degrees += 10;
     } while (degrees <= 360);
```

for loop:

```
     for (int degrees=0; degrees<=360; degrees+=10)
     {
          radians = degrees*PI/180;
        cout << degrees << " " << radians << endl;
     }
```

```
break statement:
    break;
continue statement:
    continue;
File open statements:
    sensor1.open("sensor1.dat", ios::in);
    balloon.open("balloon.dat", ios::out);
File input statement:
    sensor1 >> time >> motion;
File output statement:
    balloon << time << height << velocity/3600 << endl;
File close statement:
    sensor1.close();
```

DEBUGGING NOTES

1. Be sure to use the relational operator == instead of = in a condition for equality.

2. Put the braces surrounding a block of statements on lines by themselves; this will help you avoid omitting one of them.

3. Do not use the equality operator with floating-point values; instead, test for values "close to" a desired value.

4. Recompile your program frequently when correcting syntax errors; correcting one error may remove many error messages.

5. Use the cout statement to give memory snapshots of the values of key variables when debugging loops.

6. It is easier than you think to generate an infinite loop; be sure you know the special characters needed to abort the execution of a program on your system if it goes into an infinite loop.

KEY TERMS

block	flowchart	repetition
compound statement	for loop	selection
condition	infinite loop	sequence
data file	iteration	sentinel signal
decomposition outline	logical operator	stepwise refinement
divide and conquer	loop	top-down design
empty statement	loop-control variable	
end-of-file indicator	pseudocode	

Problems

Unit Conversions. The following problems generate tables of unit conversions. Include a table heading and column headings for the tables. Choose the number of decimal places based on the values to be printed.

1. Generate a table of conversions from degrees to radians. The first line should contain the value for 0°, and the last line should contain the values for 360°. Allow the user to enter the increment to use between lines in the table.

2. Generate a table of conversions from mi/hr to ft/s. Start the mi/hr column at 0, and increment by 5 mi/hr. The last line should contain the value 65 mi/hr. (Recall that 1 mi = 5280 ft.)

Currency Conversions. The following problems generate tables of currency conversions. Use title and column headings. Assume the following conversion rates:

1 dollar ($) = 5.045 francs (Fr)

1 yen (Y) = $ 0.010239

1 dollar ($) = 1.4685 deutsche marks (DM)

3. Generate a table of conversions from francs to dollars. Start the francs column at 5 Fr, and increment by 5 Fr. Print 25 lines in the table.

4. Generate a table of conversions from yen to deutsche marks. Start the yen column at 100 Y and print 25 lines, with the final line containing the value 10000 Y.

Temperature Conversions. The following problems generate temperature conversion tables. Use the following equations that give relationships between temperatures in degrees Fahrenheit (T_F), degrees Celsius (T_C), degrees Kelvin (T_K), and degrees Rankin (T_R):

$$T_F = T_R - 459.67° \text{ R}$$
$$T_F = \left(\tfrac{9}{5}\right) T_C + 32° \text{ F}$$
$$T_R = \left(\tfrac{9}{5}\right) T_K$$

5. Write a program to generate a table of conversions from Fahrenheit to Kelvin for values from 0°F to 200°F. Allow the user to enter the increment in degrees Fahrenheit between lines.

6. Write a program to generate a table of conversions from Celsius to Rankin. Allow the user to enter the starting temperature and increment between lines. Print 25 lines in the table.

Suture Packaging. Sutures are strands or fibers used to sew living tissue together after an injury or an operation. Packages of sutures must be sealed carefully before they are shipped to hospitals so that contaminants cannot enter the packages. The object that seals the package is referred to as a sealing die. Generally, sealing dies are heated with an electric heater. For the sealing process to be a success, the sealing die is maintained at an established temperature and must contact the package with a predetermined pressure for an established time period. The time period in which the sealing die contacts the package is called the dwell time. Assume that the acceptable range of parameters for an acceptable seal are the following:

Temperature:	150–170° C
Pressure:	60–70 psi
Dwell time:	2–2.5 s

7. A data file named `suture.dat` contains information on batches of sutures that have been rejected during a one-week period. Each line in the data file contains the batch number, the temperature, the pressure, and the dwell time for a rejected batch. The quality control engineer would like to analyze this information and needs a report that computes the percent of the batches rejected due to

temperature, the percent rejected due to pressure, and the percent rejected due to dwell time. It is possible that a specific batch may have been rejected for more than one reason and should be counted in all applicable totals. Write a program to compute and print these three percentages. Use the following test data:

Batch Number	Temperature	Pressure	Dwell Time
24551	145.5	62.3	2.23
24582	153.7	63.2	2.52
26553	160.3	58.9	2.51
26623	159.5	58.9	2.01
26642	160.3	61.2	1.98

8. Write a program to read the data file suture.dat from problem 7. Make sure that the information relates only to batches that should have been rejected. If any batch should not be in the data file, print an appropriate message with the batch information. Test your program with the specified data file; then insert records with information that should not have been rejected to be sure your program will identify these records.

Timber Regrowth. A problem in timber management is to determine how much of an area to leave uncut so that the harvested area is reforested in a certain period of time. It is assumed that reforestation takes place at a known rate per year, depending on climate and soil conditions. A reforestation equation expresses this growth as a function of the amount of timber standing and the reforestation rate. For example, if 100 acres are left standing after harvesting and the reforestation rate is 0.05, then $100 + 0.05 \times 100$, or 105 acres, are forested at the end of the first year. At the end of the second year, the number of acres forested is $105 + 0.05 \times 105$, or 110.25 acres.

9. Assume that there are 14,000 acres total with 2500 acres uncut and that the reforestation rate is 0.02. Print a table showing the number of acres reforested at the end of each year, for a total of 20 years.

10. Modify the program developed in problem 9 so that the user can enter a number of acres and the program will determine how many years are required for that number of acres to be completely reforested.

4

Modular Programming with Functions

GRAND CHALLENGE: ENHANCED OIL AND GAS RECOVERY

The design and construction of the Alaskan pipeline presented numerous engineering challenges. One of the important problems that had to be addressed was protecting the permafrost (the permanently frozen subsoil in arctic or subarctic regions) from the heat of the pipeline itself. The oil flowing in the pipeline is warmed by pumping stations and by friction from the walls of the pipe, so the supports holding the pipeline must be insulated or even cooled to keep them from melting the permafrost at their bases. In addition, the components of the pipeline haved to be very reliable, because of the inaccessibility of some locations. More importantly, component failure could cause damage to human life, animal life, and the environment around the pipeline.

SECTIONS

OBJECTIVES

In this chapter, you will

- Re-encounter the importance of dividing programs into components, or modules, that perform specific operations.
- Learn to write programmer-defined functions.
- Encounter some guidelines for using parameters and variables with functions.

4.1 MODULARITY

The execution of a C++ program begins with the statements in the `main` function. A program may also contain other functions, and it may refer to functions in another file or in a library. These functions, or **modules**, are sets of statements that typically perform an operation or that compute a value. For example, the printf function prints a line of information on the terminal screen, and the `sqrt` function computes the square root of a value.

To maintain simplicity and readability in longer and more complex problem solutions, we develop programs that use a `main` function plus additional functions, instead of using one long `main` function. By separating a solution into a group of modules, each module is simpler and easier to understand, thus adhering to the basic guidelines of structured programming presented in Chapter 3.

The process of developing a problem solution is often one of "divide and conquer," as was discussed when we first discussed the decomposition outline. Because the decomposition outline is a set of sequentially executed steps that solve the problem, it provides a good starting point for selecting potential functions. In fact, it is not uncommon for each step in the decomposition outline to correspond to one or more function references in the `main` function.

Breaking a problem solution into a set of modules has many advantages. Because a module has a specific purpose, it can be written and tested separately from the rest of the problem solution. An individual module is smaller than the complete solution, so testing it is easier. And, once a module has been carefully tested, it can be used in new problem solutions without being retested.

For example, suppose that a module is developed to find the average of a group of values. Once this module is written and tested, it can be used in other programs that need to compute an average. This **reusability** is a very important issue in the development of large software systems, because it can save development time. In fact, libraries of commonly used modules (such as the Standard C++ library) are often available on computer systems. The use of modules (called **modularity**) often reduces the overall length of a program, because many problem solutions include steps that are repeated several places in the program. By incorporating these steps that are repeated in a function, the steps can be referenced with a single statement each time they are needed.

Several programmers can work on the same project if it is separated into modules, because the individual modules can be developed and tested independently of each other. This allows the development schedule to be accelerated because some of the work can be done in parallel.

The use of modules that have been written to accomplish specific tasks supports the concept of **abstraction.** The modules contain the details of the tasks, and the programmer can reference the modules without worrying about these details. The I/O diagrams that we use in developing a problem solution are an example of abstraction—we specify the input information and the output information without giving the details of how the output information is determined.

In a similar way, we can think of modules as "black boxes" that have a specified input and compute specified information. We can use these modules to help develop a solution; thus, we are able to operate at a higher level of abstraction to solve problems. For example, the Standard C++ library contains functions that compute the logarithms of values. We can reference these functions without being concerned about the specific details, such as whether the functions are using infinite series approximations or look-up tables to compute the specified logarithms. By using abstraction, we can reduce the development time of software as we increase its quality.

To summarize, some of the advantages of using modules in a problem solution are the following:

- A module can be written and tested separately from other parts of the solution; thus, module development can be done in parallel for large projects.
- A module is a small part of the solution; thus, testing it separately is easier.
- Once a module is tested carefully, it does not need to be retested before it can be used in new problem solutions.

- The use of modules usually reduces the length of a program, making it more readable.
- The use of modules promotes the concept of abstraction, which allows the programmer to "hide" the details in modules. This allows us to use modules in a functional sense without being concerned about the specific details.

Additional benefits of modules will be pointed out as we progress through this chapter. As we begin to develop solutions to more complicated problems, the programs become longer. Therefore, we include here three suggestions for debugging longer programs. First, it is sometimes helpful to run a program using a different compiler because different compilers have different error messages. In fact some compilers have extensive error messages, whereas others give very little information about some errors. Another useful step in debugging a long program is to add comment indicators (// or /* and */) around some sections of the code so that you can focus on other parts of the program. Of course, you must be careful that you do not comment out statements that affect variables needed for the parts of the program that you want to test. Finally, test complicated functions by themselves. This is usually done with a special program called a **driver**, whose purpose is to provide a simple interface between you and the function that you are testing. Typically, this program asks you to enter the parameters that you want passed to the function, and it then prints the value returned by the function.

4.2 PROGRAMMER-DEFINED FUNCTIONS

The execution of a program always begins with the `main` function. Additional functions are called, or **invoked**, when the program encounters function names. These additional functions must be defined in the file containing the `main` function or in another available file or library of files. (If the function is included in a system library file, such as the Standard C++ library, it is often called a **library function**; other functions are usually called **programmer-written** or **programmer-defined functions**.) After executing the statements in a function, the program execution continues with the statement that called the function.

Function Definition

The sinc(x) function, plotted in Figure 4.1, is commonly used in many engineering applications. The most common definition for sinc(x) is the following:

$$f(x) \; = \; \frac{\sin(x)}{x}$$

(The sinc(x) function is also occasionally defined as $\sin(\pi x)/\pi x$.) The values of this function can be easily computed except for sinc(0), which gives an indeterminant form of 0/0. In this case, L'Hopital's theorem from calculus can be used to prove that sinc(0) = 1.

Assume that you want to develop a program that allows the user to enter interval limits, a and b. The program should then compute and print 21 values of sinc(x) for values of x evenly spaced between a and b, inclusively. Thus, the first value of x should be a. An increment should then be added to obtain the next value of x, and so on, until the 21st value, which should be b. Therefore, the increment in x is

$$x\_increment \; = \; \frac{\text{interval width}}{20}$$

$$= \; \frac{b - a}{20}$$

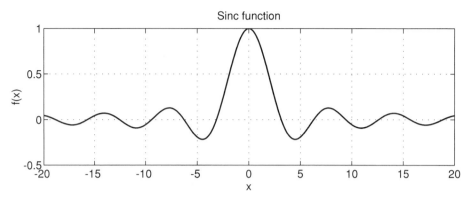

Figure 4.1. Sinc function in $[-20, 20]$.

Select values for a and b, and convince yourself that, with this increment, and with a as the first value, the 21st value will be b.

Because sinc(x) is not one of the mathematical functions provided by the Standard C++ library, we implement this problem solution two ways. In one solution we include the statements to perform the computations of sinc(x) in the main function; in the other solution we write a programmer-defined function to compute sinc(x) and then reference the programmer-defined function each time that the computations are needed. Both solutions are now presented so that you can compare them.

Solution 1

```
//-----------------------------------------------------------
//   Program chapter4_1
//   /
//   This program prints 21 values of the sinc
//   function in the interval [a,b] using
//    computations within the main function.

#include <iostream.h>
#include <iomanip.h>
#include <stdlib.h>
#include <math.h>

int main()
{
   //  Define variables.
   double a, b, new_x, sinc_x;

   //  Get interval endpoints from the user.
   cout << "Enter endpoints a and b (a<b): " << endl;
  cin >> a >> b;
   double x_incr = (b - a)/20;

   //  Compute and print table of sinc(x) values.
   cout << "x and sinc(x) " << endl;
  for (int k=0; k<=20; k++)
  {
       new_x = a + k*x_incr;
       if (fabs(new_x) < 0.0001)
         sinc_x = 1.0;
      else
            sinc_x = sin(new_x)/new_x;
```

```
            cout << setw(4) << new_x << "   "
                    << setprecision(4) << sinc_x << endl;
   }

      //  Exit program.
      return EXIT_SUCCESS;
}
//------------------------------------------------------------
```

We now present a second solution that uses a programmer-defined function to compute values of sinc(x).

Solution 2

```
//------------------------------------------------------------
//  Program chapter4_2
//
//  This program prints 21 values of the sinc
//  function in the interval [a,b] using a
//  programmer-defined function.

#include <iostream.h>
#include <iomanip.h>
#include <stdlib.h>
#include <math.h>

int main()
{
   //  Define variables and function prototype.
  double a, b, new_x;
   double sinc(double x);

   //  Get interval endpoints from the user.
   cout << "Enter endpoints a and b (a<b): " << endl;
  cin >> a >> b;
   double x_incr = (b - a)/20;
   //  Compute and print table of sinc(x) values.
   cout << "x and sinc(x) " << endl;
  for (int k=0; k<=20; k++)
  {
      new_x = a + k*x_incr;
      cout << setw(4) << new_x << "   "
              << setprecision(4) << sinc(new_x) << endl;
   }
   //  Exit program.
   return EXIT_SUCCESS;
}
//------------------------------------------------------------
//  This function evaluates the sinc function.
//
double sinc(double x)
{
   if (fabs(x) < 0.0001)
      return 1.0;
   else
      return sin(x)/x;
}
//------------------------------------------------------------
```

The following output represents an example interaction that could occur with either program:

```
Enter endpoints a and b (a<b):
-5 5
x and sinc(x)
   -5   -0.1918
 -4.5   -0.2172
   -4   -0.1892
 -3.5   -0.1002
   -3    0.047
 -2.5    0.2394
   -2    0.4546
 -1.5    0.665
   -1    0.8415
 -0.5    0.9589
    0    1
  0.5    0.9589
    1    0.8415
  1.5    0.665
    2    0.4546
  2.5    0.2394
    3    0.047
  3.5   -0.1002
    4   -0.1892
  4.5   -0.2172
    5   -0.1918
```

Figure 4.2 contains plots of the 21 values computed for four different intervals $[a, b]$. Because the program computes only 21 values, the resolution in the plots is affected by the size of the interval—a smaller interval has better resolution than a larger interval. Now that you have an example of a program with a programmer-defined function, we present a more general discussion of the statements in a function.

A function consists of a definition statement followed by declarations and statements. The first part of the definition statement defines the type of value that is returned by the function. If the function does not return a value, the type is void. The function name and parameter list follow the return_type. Thus, the general form of a function is

```
return_type function_name(parameter declarations)
{
    function body
}
```

The parameter declarations represent the information passed to the function. If there are no input parameters (also called arguments), then the parameter declarations should be void. Additional variables used by a function are defined in the declarations. The declarations and the statements within a function are enclosed in braces. The function name should be selected to help document the purpose of the function. You should also include comments within the function to further describe the purpose of the function and to document the steps. We also use a comment line with dashes to separate a programmer-defined function from the main function and from other programmer-defined functions.

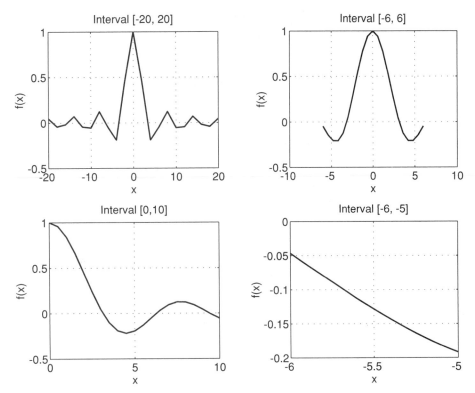

Figure 4.2. Program output for four intervals.

All functions should include a `return` statement, which has the following general form:

```
return expression;
```

The expression specifies the value to be returned to the statement that referenced the function. The expression type should match the `return_type` indicated in the function definition to avoid potential errors. The cast operator (discussed in Chapter 2) can be used to explicitly specify the type of the expression if necessary. A `void` **function** does not return a value and thus has this general definition statement:

```
void function_name(parameter declarations)
```

The `return` statement in a `void` function does not contain an expression and has this form:

```
return;
```

Compare the general form of a function that has just been described with the `sinc` function defined in program `chapter 4_2`. Also, note that the `main` function of this solution is easier to read because it is shorter than the `main` function in the first solution.

Functions can be defined before or after the `main` function. (Remember that a right brace specifies the end of the `main` function.) However, one function must be completely defined before another function begins; function definitions cannot be nested within each other. In our programs, we include the main function first; then additional functions are included in the order in which they are referenced in the program.

We now take a closer look at the interaction between a statement that references a function and the function itself.

Function Prototype

The main function presented in program chapter4_2 contained the following statement in its declarations:

```
double sinc(double x);
```

This statement is a **function prototype** statement. It informs the compiler that the main function will reference a function named sinc, that the sinc function expects a double parameter, and that the sinc function returns a double value. The identifier x is not being defined as a variable; it is just used to indicate that a value is expected as an argument by the sinc function. In fact, it is valid to include only the argument types in the **function prototype** statement:

```
double sinc(double);
```

Both of these prototype statements give the same information to the compiler. We recommend using parameter identifiers in prototype statements because the identifiers help document the order and definition of the parameters.

A function prototype can be included with preprocessor directives, or because a function prototype is defining the type of value being returned by the function, it can also be included with other variable declarations. For example, the declarations of program chapter4_2 are

```
//  Define variables and function prototype.
double a, b, new_x;
double sinc(double x);
```

These statements could also have been written in the following form:

```
//  Define variables and function prototype.
double a, b, new_x, sinc(double x);
```

In the programs in this book, we list function prototypes are listed on separate declaration statements to make it easier to identify them.

Function prototype statements should be included for all functions referenced in a program. Header files, such as iostream.h and math.h, contain the prototype statements for many of the functions in the Standard C++ library. Otherwise, you would need to include individual prototype statements for functions such as setw and sqrt in your programs. If a programmer-defined function references other programmer-defined functions, it will also need additional prototype statements.

If a program references a large number of programmer-defined functions, it becomes cumbersome to include all the function prototype statements. In these cases, a custom header file can be defined that contains the function prototypes and any related symbolic constants. A header file must have a filename that ends with a suffix of .h. The file is then referenced with an include statement, using double quotes around the filename. Custom header files are often used to accompany routines that are shared by programmers.

Parameter List

The definition statement of a function defines the parameters that the function requires; these are called **formal parameters**. Any statement that references the function must include values that correspond to the parameters; these are called **actual**

parameters. For example, consider the `sinc` function developed earlier in this section. The definition statement of this function is

```
double sinc(double x)
```

and the statement from the `main` program that references the function is

```
cout << new_x << "   " << sinc(new_x);
```

Thus, the variable `x` is the formal parameter, and the variable `new_x` is the actual parameter. When the reference to the `sinc` function in the `cout` statement is executed, the value in the actual parameter is copied to the formal parameter, and the steps in the `sinc` function are executed using the new value in `x`. The value returned by the `sinc` function is then printed.

It is important to note that the value in the formal parameter is not moved back to the actual parameter when the function is completed. We illustrate these steps with a memory snapshot that shows the transfer of the value from the actual parameter to the formal parameter, assuming that the value of `new_x` is 5.0:

ACTUAL PARAMETER   **FORMAL PARAMETER**

new_x   `5.0`   → x   `5.0`

After the value in the actual parameter is copied to the formal parameter, the steps in the `sinc` function are executed. When debugging a function, it is a good idea to use `printf` statements to provide a memory snapshot of the actual parameters before the function is referenced, and of the formal parameters at the beginning of the function.

Valid references to the `sinc` function can also include expressions and can include other function references, as shown in these example references to the `sinc` function:

```
cout << sinc(x+2.5) << endl;

cin >> y;
cout << sinc(y) << endl;

z = x*x + sinc(2*x);

w = sinc(fabs(y));
```

In all of these example references, the formal parameter is still `x`, but the actual parameter is `x + 2.5`, or `y`, or `2*x`, or `fabs(y)`, depending on the reference selected.

If a function has more than one parameter, the formal parameters and the actual parameters must match in number, type, and order. A mismatch between the number of formal parameters and actual parameters can be detected by the compiler using the function prototype statement. If the type of an actual parameter is not the same as the corresponding formal parameter, the value of the actual parameter will be converted to the appropriate type. This conversion is called **coercion of arguments**, and may or may not cause errors. Recall that converting values to a higher type (such as from `float` to `double`) generally works correctly. Converting values to a lower type (such as from `float` to `int`), on the other hand, often introduces errors.

Additional errors can be introduced if the actual parameters are out of order. These errors may not be detected by the compiler and can be difficult to detect. Therefore, be especially careful that the order of the formal parameters and the actual parameters match.

The function reference in the `sinc` example is a **call-by-value** reference, or a **reference by value**. In general, a C++ function cannot change the value of an actual parameter. An exception occurs when an actual parameter *is* an array; this will be discussed in the next chapter.

PRACTICE!

Consider the following function:

```
/*------------------------------------------------------------*/
int positive(double a, double b, double c)
{
   int count = 0;

   if (a >= 0)
      count++;
   if (b >= 0)
      count++;
   if (c >= 0)
      count++;

   return count;
}
/*------------------------------------------------------------*/
```

Assume that the function is referenced with the following statements:

```
x = 25;
total = positive(x, sqrt(x), x-30);
```

1. Show the memory snapshot of the actual parameters and the formal parameters.
2. What is the new value of `total`?

Storage Class and Scope

The example programs presented so far, have declared variables within a `main` function and within programmer-defined functions. It is also valid to define a variable before the `main` function. Therefore, it is important to be able to determine the scope of a function or a variable. **Scope** refers to the portion of the program in which it is valid to reference the function or variable. Scope is also sometimes defined in terms of the portion of the program in which the function or variable is visible or accessible. The scope of a variable is directly related to its **storage class**, which can be automatic, external, or static.

Before describing storage classes, a distinction needs to be made between local and global variables. **Local variables** are defined within a function, and include the formal parameters and any other variables declared in the function. A local variable can be accessed only in the function that defined it. A local variable has a value when its function is being executed, but its value is not retained when the function is completed. **Global variables** are defined outside the `main` function and other programmer-defined functions, so they can be accessed by any function within the program. However, to reference a global or an external variable, a declaration within the function must include the keyword `extern` before the type designation to tell the computer to look outside the function for the variable.

The **automatic storage class** is used to represent local variables. This is the default storage class, but it can also be specified with the keyword auto before the type designation. The **external storage class** is used to represent global variables; the extern designation must be used within functions; it is optional in the original definition of a global variable.

The memory assigned to an external variable is retained for the duration of the program. Although an external variable can be referenced from a function using the proper declaration, using global variables is generally discouraged. In general, parameters are preferred for transferring information to a function because the parameter is evident in the function prototype, whereas the external variable is not visible in the function prototype.

Function names also have external storage class and can be referenced from other functions. Function prototypes included outside of any function are also external references, and are available to all other functions in the program. This explains why we do not need to include `math.h` in every function that references a mathematical function. However, the parameter variables in the function prototype are known only in the function prototype statement.

The **static storage class** is used to specify that the memory for a variable should be retained during the entire program execution. Therefore, if a local variable in a function is given a static storage class assignment by using the keyword `static` before its type specification, the variable will not lose its value when the program exits the function in which it is defined. A `static` variable could be used to count the number of times that a function was invoked, because the value of the count would be preserved from one function call to another.

C++ STATEMENT SUMMARY

Function definition

```
return_type function_name(parameter types)
{
    function body
}
```

Return statement

```
return;
return (a + b)/2;
```

Function Prototype

```
double sinc(double x);
double sinc(double);
void print_data(int x, int y);
```

Debugging Notes

1. If you are having difficulty understanding the error messages from a compiler, try running the program on another compiler to obtain different error messages.
2. When debugging a long program, add comment indicators ((`//` or `/*` and `*/`) around some sections of the code so that you can focus on other parts of the program.
3. Test a complicated function by itself, using a driver program.
4. Make sure that the value returned from a function matches the function `return_type`. If necessary, use the cast operator to convert a value to the proper type.
5. Functions can be defined before or after the `main` function, but not within it.
6. Always use function prototype statements to avoid errors in parameter passing.

7. Use `cout` statements to generate memory snapshots of the actual parameters before a function is referenced and of the formal parameters at the beginning of the function.

8. Carefully match the type, order, and number of actual parameters with the formal parameters of a function.

KEY TERMS

abstraction
actual parameter
automatic class
call by value
coercion of arguments
driver
external class

formal parameter
function prototype
global variable
invoke
library function
local variable
modularity

module
programmer-defined function
reusability
static class
scope
void functions

Problems

Print Routines. The following problems develop functions that are useful when printing reports. These are all void functions because they print information, but they do not return any values.

1. Write and test a function that will print your name, course title, and homework number in the following format:

```
Joey Smith
 Engineering 101
Homework #5
```

Your name and course title should be included within the output statements; the homework number is an input parameter. Assume that the function prototype is

```
void header(int hw_number);
```

2. Write and test a function that prints two totals in the following format:

```
Summary Information:
    Total 1 xxxx.xx
    Total 2 xxxx.xx
      Combined Totals     xxxxx.xx
```

Assume that the function prototype is

```
void summary(double total_1, double total_2);
```

3. Write and test a function that prints the following error message:

```
Error occurred in the processing of the data file.
Recheck the data before rerunning this program.
```

Assume that the function prototype is

```
void error(void);
```

4. Write and test a function that prints an error message in the following format:

```
Errors occurred in processing the data file.
xxx values were out of the appropriate range.
```

The number of values out of range is an input parameter to the function. Assume that the function prototype is

```
void error_total(int error_count);
```

5. Write and test a function that prints an error message that depends on the value in an integer parameter. If the value of the error flag is 1, print

    ```
    Error identified in the distance data.
    ```

    If the value of the error flag is 2, print

    ```
    Error identified in the velocity data.
    ```

    If the value of the error flag is 3, print

    ```
    Error identified in the acceleration data.
    ```

    For all other values of the error flag, print

    ```
    Unidentified error occurred.
    ```

    Assume that the function prototype is

    ```
    void error_message(int error_flag);
    ```

Function Evaluations. Functions are often used to evaluate a mathematical function that is not included in the Standard C++ library. The following problems develop functions to return function values.

6. Write and test a function that receives three integers. The function should return the maximum value. Assume that the function prototype is

    ```
    int maximum_int(int a, int b, int c);
    ```

7. Write and test a function that returns the value of f(x) where

$$f(x) \ = \ \frac{3x^2 + x - 2}{x^2 - 2x + 1}$$

    Assume that the function prototype is

    ```
    double f(double x);
    ```

8. Write and test a function that returns the value of g(x) where

$$g(x) \ = \ 0 \text{ , if } x < 0$$
$$= \ 6e^{(x-3)} \text{ , if } x \geq 0$$

    Assume that the function prototype is

    ```
    double g(double x);
    ```

9. Wri4te and test a function that returns the radius of the line from a point (x,y) on a plane to the origin, where

$$\text{radius} \ = \ \sqrt{x^2 + y^2}$$

    Assume that the function prototype is

    ```
    double length(double x, double y);
    ```

10. Write and test a function that returns the quadrant number of a point (x,y) on a plane. Recall that points in quadrant 1 have positive x and y values, points in quadrant 2 have a negative x value and a positive y value, points in quadrant 3 have negative x and y values, and the remaining points are in quadrant 4. If a point is on an axis, choose the quadrant with the lower quadrant number. Assume that the function prototype is

    ```
    int quadrant(double x, double y);
    ```

5

One-Dimensional Arrays

GRAND CHALLENGE:
SPEECH RECOGNITION

The modern jet cockpit has literally hundreds of switches and gauges. Several research programs are investigating the feasibility of using a speech-recognition system in the cockpit to serve as a pilot's assistant. The system would respond to verbal requests from the pilot for information such as fuel status or altitude. The pilot would use words from a small vocabulary that the computer had been trained to understand. In addition to understanding a specific vocabulary, the system would also have to be trained using the speech for the pilot who would be using the system. This training information could be stored on a diskette and inserted into the onboard computer at the beginning of a flight so the system could recognize the current pilot. The computer system would also use speech synthesis to respond to the pilot's request for information.

OBJECTIVES

In this chapter, you will

- Learn about the array, a data structure frequently used in solving engineering and scientific problems.
- Define, initialize, and perform computations with arrays.
- Use arrays in input and output statements and as function arguments.

5.1 ARRAY DEFINITIONS AND COMPUTATIONS

When solving engineering problems, it is important to be able to visualize the data related to the problem. Sometimes the data is just a single number, such as the radius of a circle. Other times the data may be coordinates in a plane that can be represented as a pair of numbers, with one number representing the x-coordinate and the other number representing the y-coordinate. There are also times when we want to work with a set of similar data values, but we do not want to give each value a separate name. For example, suppose that we have a set of 100 temperature measurements that we want to use to perform several computations. Obviously, we do not want to use 100 different names for the temperature measurements, so we need a method for working with a group of values using a single identifier. One solution to this problem uses a data structure called an array. A **one-dimensional** array can be visualized as a list of values arranged in either a row or a column, as shown below:

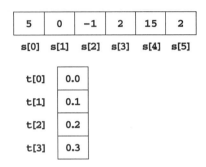

We assign an identifier to an array, and then distinguish between **elements** or values in the array using **subscripts**. In C++, the subscripts always start with 0 and increment by 1. Thus, using the example arrays, the first value in the s array is referenced by s[0], and the third value in the t array is referenced by t[2].

Arrays are convenient for storing and handling large amounts of data, so there is a tendency to use them in algorithms when they are not necessary. Arrays are more complicated to use than simple variables and make programs longer and more difficult to debug. Therefore, use arrays only when it is necessary to have the complete set of data available in memory.

Definition and Initialization

An array is defined using declaration statements. An integer expression in brackets follows the identifier and specifies the number of elements in the array. Note that all elements in an array must be the same data type. The declaration statements for the two example arrays are as follows:

```
int s[6];
double t[4];
```

An array can be initialized with declaration statements or with program statements. To initialize the array with the declaration statement, the values are specified in a sequence that is separated by commas and enclosed in braces. The following statements define and initialize the example arrays s and t:

```
int s[6] = {5, 0, -1, 2, 15, 2};
double t[4] = {0.0, 0.1, 0.2, 0.3};
```

If the initializing sequence is shorter than the array, the rest of the values are initialized to zero. Hence, the following statement defines an integer array of 100 values; each value is also initialized to zero:

```
int s[100] = {0};
```

If an array is specified without a size but with an initialization sequence, the size is defined to be equal to the number of value in the sequence. Thus, the following statements also define the arrays s and t:

```
int s[] = {5, 0, -1, 2, 15, 2};
double t[] = {0.0, 0.1, 0.2, 0.3};
```

The size of an array must be specified in the declaration statement, using either a constant within brackets or by an initialization sequence within braces.

You can also initialize arrays with program statements. For example, suppose that you want to fill a `double` array g with the values 0.0, 0.5, 1.0, 1.5,..., 10.0. Because there are 21 values, listing the values on the declaration statement would be tedious. Instead, use the following statements to define and initialize this array:

```
//  Define variable.
double g[21];
 .
 .
 .
//  Assign initial values to the array g.
for (int k=0; k<21; k++)
   g[k] = k*0.5;
```

It is important to recognize that the condition in this `for` statement must specify a final subscript value of 20, and not 21. It is a common mistake to specify a subscript that is one value more than the largest valid subscript. This error can be very difficult to find because it accesses values outside the array. Because this error is generally not detected during the program execution, it is important to be careful about exceeding the array subscripts. In the programs in this book, loop conditions are selected that specifically use the final value as a reminder to carefully write the condition to avoid errors. That is why this example uses the condition `k<=20` instead of `k<21`, although both work properly. Also, this book generally uses k as the subscript for a one-dimensional array.

Arrays are often used to store data read from files. An example might be a data file named `sensor3.dat` that contains ten time and motion measurements collected from a seismometer. To read these values into arrays named `time` and `motion`, you could use these statements:

```
//  Define variables.
double time[10], motion[10];
 .
 .
 .
//  Open file and read data into arrays.
fstream sensor3;
sensor3.open("sensor3.dat", ios::in);
for (int k=0; k<10; k++)
    sensor3 >> time[k] >> motion[k];
```

PRACTICE!

Show the contents of the arrays defined in each of the following sets of statements.

1. `int x[10] = {-5, 4, 3};`

2. `double z[4];`

   ```
   ...
   z[1] = -5.5;
   z[2] = z[3] = fabs(z[1]);
   ```

PRACTICE!

```
3. int k, time[9];
   ...
   for (k = 0; k <= 8; k++)
     time[k] = k - 4)*0.1;
```

5.1.1 Computations and I/O

You perform computations with array elements just like with simple variables, except that you have to use a subscript to specify an individual array element. To illustrate, the following program reads an array y of 100 floating-point values from a data file. The program determines the average value of the array and stores it in y_ave. Then, the number of values in the array y that are greater than the average are counted and printed.

```
//------------------------------------------------------------
//  Program chapter5_1
// /
//  This program reads 100 values from a data file
//  and determines the number of values greater
//  than the average.

#include <iostream.h>
#include <fstream.h>
#include <stdlib.h>

int main()
{
   // Define constant and variables.
   const int N=100;
   double y[N], sum=0;

   // Open file, read data into an array,
   // and compute a sum of the values.
   fstream lab;
    lab.open("lab1.dat", ios::in);
   for (int k=0; k<N; k++)
{
   lab >> y[k];
   sum += y[k];
}

   // Compute average and count values that
   // are greater than the average.
   int count=0;
    double y_ave=sum/N;
  for (k=0; k<N; k++)
      if (y[k] > y_ave)
         count++;

   // Print count.
   cout << count << " values greater than the "
      << "average value" << endl;

   // Close file and exit program.
   lab.close();
   return EXIT_SUCCESS;
}
//------------------------------------------------------------
```

If the purpose of this program had been to determine the average of the values in the data file, an array would not have been necessary. The loop to read values could read each value into the same variable, adding its value to a sum before the next value is read. However, because we needed to compare each value to the average in order to count the number of values greater than the average, an array was needed so that we could access each value again.

Array values are printed using a subscript to specify the individual value desired. For example, the following statement prints the first and last values of the array y used in the previous example:

```
cout << "first and last array values:" << endl;
    << y[0] << "  " << y[N-1] << endl;
```

The following loop prints all 100 values of y, one per line:

```
cout << "y values: " << endl;
for (int k=0; k<N; k++)
   cout << y[k] << endl;
```

When printing a large array, such as this one, you would probably like to print several numbers on the same line. The following statements use the modulus operator to skip to a new line before each group of five values is printed:

```
cout << "y values:" << endl;
for (int k=0; k<N; k++)
   if (k%5 == 0)
       cout << endl << y[k];
   else
      cout << y[k];
cout << endl;
```

You can use statements similar to the ones illustrated here to write array values to a data file. For example, the following statement will print the value of y[k] on a line in a data file with a file pointer `sensor`:

```
sensor << y[k] << endl;
```

The new line indicator is included, so the next value written to the file will be on a new line.

The number of elements in an array is used in the array declaration and in loops used to access the elements in the array. If the number of elements is changed, there are several places in the program that need to be modified. Changing the size of an array is simplified if a symbolic constant is used to specify the size of the array. Then, to change the size, only the preprocessor directive needs to be changed. This style suggestion is especially important in programs that contain many modules or in programming environments in which several programmers are working on the same software project.

Table 5-1 gives an updated precedence order that includes subscript brackets. Brackets and parentheses are associated before the other operators. If parentheses and brackets are nested, the innermost set is evaluated first.

TABLE 5-1   Operator Precedence

PRECEDENCE	OPERATION	ASSOCIATIVITY
1	( ) [ ]	innermost first
2	+ - ++ -- (type)!	right to left
3	* / %	left to right
4	+ -	left to right
5	< <= > >=	left to right
6	== !=	left to right
7	&&	left to right
8	//	left to right
9	= += -= *= /= %=	right to left

PRACTICE!

Assume that the variable k and the array s have been defined with the following statement:

```
int, s[]={3, 8, 15, 21, 30, 41};
```

Give the output for each of the following sets of statements.

1.
```
for (int k=0; k<6; k+=2)
    cout << s[k] << "  " << s[k+1] << endl;
```
2.
```
for (int k=0; k<6; k++)
    if (s[k]%2 == 0)
        cout << s[k] << endl;
```

5.2 ARRAYS AS FUNCTION ARGUMENTS

When the information in an array is passed to a function, two parameters are usually used. One parameter specifies the specific array, and the other parameter specifies the number of elements used in the array. By specifying the number of elements of the array that are to be used, the function becomes more flexible.

For example, if the function specifies an integer array, the function can be used with any integer array. The parameter that specifies the number of elements assures that we use the correct size.

Also, the number of elements used in an array may vary from one time to another. For example, the array may use elements read from a data file. The number of elements then depends on the specific data file used when the program is run. In all of these examples, though, the array must be declared to a maximum size in the main function. Then the actual number of elements used can be less than or equal to that maximum size.

Consider the following program that reads an array from a data file and then references a function to determine the maximum value in the array. The variable npts is used to specify the number of values in the array. The value of npts can be less than or equal to the defined size of the array, which is 100. The function has two arguments—the name of the array and the number of points in the array, as indicated in the function prototype statement.

```
//---------------------------------------------------------
//  Program chapter5_2
//  /
//  This program reads values from a data file and
//   determines the maximum value with a function.
#include <iostream.h>
#include <fstream.h>
#include <stdlib.h>

int main()
{
   //  Define constant, variables, and prototype.
   const int N=100;
   int npts;
   double y[N];
   double array_max(double x[], int n);

   //  Input the number of data values.
   cout << "Enter the number of data values: ";
   cin >> npts;

   //  Open file and read data into an array.
   fstream lab;
   lab.open("lab1.dat", ios::in);
   for (int k=0; k<npts; k++)
     lab >> y[k];

   //  Find and print the maximum value.
   cout << "maximum value: "
        << array_max(y,npts) << endl;

   //  Close file and exit program.
   lab.close();
   return EXIT_SUCCESS;
}
//---------------------------------------------------------
//  This function returns the maximum
//   value in the array x with n elements.
//  /
double array_max(double x[], int n)
{
   //  Define variable.
   double max_x;

   //  Determine the maximum value in the array.
   max_x = x[0];
   for (int k=1; k<n; k++)
      if (x[k] > max_x)
         max_x = x[k];

   //  Return maximum value.
   return max_x;
}
//---------------------------------------------------------
```

This program assumes that there will not be more than 100 values in the file; otherwise, this program will not work correctly. Arrays must be specified to be as large as, or larger than, the maximum number of values to be read into them.

The purpose of program chapter5_2 was to illustrate the use of an array as a function argument. If the purpose of this program had been to determine the maximum of the data values in the file, an array would not have been necessary; the maximum could have been determined as the data values were read.

Call-by-Address References

There is a very significant difference in using arrays as parameters and in using simple variables as parameters. When a simple variable is used as a parameter, the value is passed to the formal parameter in the function, and the value of the original value cannot be changed; this is a **call-by-value** reference. When an array is used as a parameter, the memory address of the array is passed to the function instead of the entire set of values in the array. Therefore, the function references values in the original array; this is a **call-by-address** reference. Because a function accesses the original array values, we must be very careful that we do not inadvertently change values in an array within a function. Of course, there may also be occasions when we wish to change the values in the array.

PRACTICE!

Assume that the following variables are defined:

```
int k=6;

double data[]={1.5, 3.2, -6.1, 9.8, 8.7, 5.2};
```

Give the values of the following expressions that reference the array_max function presented in this section.

1. `array_max(data,6);`

2. `array_max(data,5);`

3. `array_max(data,k - 3);`

4. `array_max(data,k%5);`

Statistical Measurements

Analyzing data collected from engineering experiments is an important part of evaluating the experiments. This analysis ranges from simple computations on the data, such as calculating the average value, to more complicated analyses. Many of the computations or measurements using data are statistical measurements because they have statistical properties that change from one set of data to another. For example, the sine of 60° is an exact value that is the same value every time we compute it, but the number of miles to the gallon that we get with our car is a statistical measurement because it varies somewhat depending on parameters such as the temperature, the speed that we travel, the type of road, and whether we are in the mountains or the desert.

When evaluating a set of experimental data, we often compute the maximum value, minimum value, mean or average value, and the median. This section develops functions that you can use to compute the values using an array as input. The functions will be useful in solutions to problems at the end of the chapter. It is important to note that these functions assume that there is at least one value in the array.

Maximum, Minimum   A function for determining the maximum value in an array was presented earlier in this section; a similar function can easily be developed for determining the minimum value.

Average   The Greek symbol μ (mu) is used to represent the average or **mean value,** as shown in the following equation, which uses summation notation:

$$\mu = \frac{\sum_{k=0}^{n-1} x_k}{n}$$

where $\sum_{k=0}^{n-1} x_k = x_0 + x_1 + x_2 + \ldots + x_{n-1}$. The average of a set of values is always a floating-point value, even if all the data values are integers. This function computes the mean value of a `double` array of n values:

```
//-----------------------------------------------------------
//  This function returns the average or
//  mean value of an array with n elements.
//  /
double array_mean(double x[], int n)
{
   //  Define and initialize variable.
   double sum=0;

   //  Determine mean value.
   for (int k=0; k<n; k++)
      sum += x[k];

   //  Return mean value.
   return sum/n;
}
//-----------------------------------------------------------
```

Note that the variable sum was initialized to zero on the declaration statement. It could also have been initialized to zero with an assignment statement. In either case, the value of sum is initialized to zero each time that the function is referenced.

Median   The median is the value in the middle of a group of values, assuming that the values are sorted. If there is an odd number of values, the median is the value in the middle; if there is an even number of values, the median is the average of the values in the two middle positions. For example, the median of the values {1, 6, 18, 39, 86} is the middle value, or 18; the median of the values {1, 6, 18, 39, 86, 91} is the average of the two middle values, or (18 1 39)/2 or 28.5. Assume that a group of sorted values is stored in an array and that n contains the number of values in the array. If n is odd, the subscript of the middle value can be represented by floor(n/2), as in floor(5/2), which is 2. If n is even, then the subscripts of the two middle values can be represented by floor(n/2) - 1 and floor(n/2), as in floor(6/2) - 1 and floor(6/2), which are 2 and 3 in the example above.

The following function determines the median of a set of values stored in an array. We assume that the values are sorted (into either ascending or descending order).

```
//-----------------------------------------------------------
//  This function returns the median
//  value in an array x with n elements.
//  /
double array_median(double x[], int n)
{
   //  Define variable.
   double median_x;

   //  Determine median value.
   int k = floor(n/2);
   if (n%2 != 0)
      median_x = x[k];
   else
      median_x = (x[k-1] + x[k])/2;
```

```
    //  Return median value.
    return median_x;
}
//----------------------------------------------------------
```

Go through this function by hand using the two sets of data values given in this discussion.

C++ STATEMENT SUMMARY

Array Declaration

```
int a[5], b[]={2, 3, -1};
```

Debugging Notes

1. Use arrays only when it is necessary to keep all the data available in memory.
2. Be careful not to exceed the maximum subscript value when referencing an element in an array.
3. Select conditions in `for` loops to specifically use an equality with the maximum subscript value; this helps avoid errors with subscript ranges.
4. An array must be declared to be as large as, or larger than, the maximum number of values to be stored in it.
5. Because an array reference in a function is a call-by-address reference, be careful that you do not inadvertently change values in an array in the function.

KEY TERMS

array	element	median
call-by-address	mean value	one-dimensional
call-by-value		

Problems

These problems are all stated in terms of writing a function that has an array as an argument. Alternative problems can be defined that do not require functions by changing each problem to one of reading 20 values from the keyboard and storing them in an array. The program should then determine and print the value computed as an array characteristic (problems 1-5) or modify and print the array values (problems 6-10).

Array Characteristics. It is very convenient to have a group of functions for determining information stored in an array. The function `array_max`, `array_mean`, and `array_median` developed in this chapter are examples of these types of routines. The following problems develop additional functions that are useful for working with arrays.

1. Write and test a function that will determine if the values in an array are in ascending order or descending order. The function value should be 1 if the array values are in ascending order, −1 if they are in descending order, and 0 if they are in neither order. Assume that the function prototype is the following:

```
int order(int x[], int npts);
```

2. Write and test a function that will compare values in two arrays of the same size. If the values in the two arrays are the same, the function should return a 1; otherwise, it should return 0. Assume that the function prototype is the following:

```
int identical(int a[], int b[], int npts);
```

3. Write and test a function that counts the number of values in an array that are greater than a specified target value. Assume that the function prototype is the following:

```
int greater_count(int x[], int npts, int target);
```

4. Write and test a function that counts the number of values in an array that are less than or equal to a specified target value. Use the function developed in problem 3 in your solution. Assume that the function prototype is the following:

```
int less_equal_count(int x[], int npts, int target);
```

5. Write and test a function that determines the largest difference between two adjacent values in an array. If there is only one value in the array, the difference should be zero. Assume that the function prototype is the following:

```
double diff(double y[], int npts);
```

Array Modification. The following set of functions modifies the values in an array.

6. Assume that an array contains angles in radians. Write a function that converts the angles to degrees. Assume that the function prototype is the following:

```
void rad_to_deg(double b[], int npts);
```

7. Assume that an array contains angles in degrees. Write a function that converts any angles outside of the interval [0,360] to an equivalent angle in the interval [0,360]. For example, an angle of 380.5 degrees should be converted to 20.5 degrees. Assume that the function prototype is the following:

```
void reduce_angle(double b[], int npts);
```

8. Write and test a function that will replace (or "clip") any value above a specified value with the specified value. Assume that the function prototype is the following:

```
void clip(double b[], int npts, double peak_value);
```

To illustrate, the following function reference will replace any value above 5.0 in the array x with the value 5.0:

```
clip(x, n, 5.0);
```

9. Write and test a function that will subtract the mean value of an array from each element in the array. (The mean value of these new values is then always zero.) Use the function array_mean in your function. Assume that the function prototype is the following:

```
void remove_mean(double x[], int npts);
```

10. There are a number of ways to normalize, or scale, a set of values. One common normalization technique scales the values so that the minimum value goes to 0, the maximum value goes to 1, and other values are scaled accordingly. Using this normalization, the values in the array below are normalized:

Array values:

−2	−1	2	0

Normalized array values:

0.0	0.25	1.0	0.5

The equation that computes the normalized value from a value x_k in the array is the following:

$$\text{Normalized } x_k = \frac{x_k - \min_x}{\max_x - \min_x}$$

where minx and maxx represent the minimum and maximum values in the array x. If you substitute the minimum value for xk in this equation, the numerator is zero; thus, the normalized value for the minimum is zero. If you substitute the maximum value for xk in this equation, the numerator and denominator are the same value; hence, the normalized value for the maximum is 1.0. Write and test a function that has a one-dimensional `double` array and the number of values in the array as its arguments. Normalize the values in the array using the technique presented above. Use the function `array_max` that was developed in this chapter. Assume that the function prototype is the following:

```
void norm(double x[], int npts);
```

6

Character Data

GRAND CHALLENGE: MAPPING THE HUMAN GENOME

Deciphering the human genetic code involves locating, identifying, and determining the function of each of the 50,000 to 100,000 genes that are contained in human DNA. Each gene is a double-helix strand composed of base pairs of adenine bonded with thymine, or cytosine bonded with guanine, that are arranged in a steplike manner with phosphate groups along the side. DNA directs the production of proteins, so the proteins produced by a cell provide a key to the sequence of base pairs in the DNA. Instrumentation developed for genetic engineering is extremely useful in this detective work. A protein sequencer developed in 1969 can identify the sequence of amino acids in a protein molecule. Once the amino acid order is known, biologists can begin to identify the gene that made the protein. A DNA synthesizer, developed in 1982, can build small genes or gene fragments out of DNA. This research and its associated instrumentation are key components in beginning to address the mapping of the human genome.

SECTIONS

- 6.1 Character Information
- 6.2 Character Initialization and Computations
- 6.3 Character Functions
- C++ Statement Summary
- Key Terms

OBJECTIVES

In this chapter, you will

- Learn the importance of character data in engineering and science.
- Become familiar with the ASCII code.
- Initialize, read, and print character variables.
- Learn about functions that work with characters.

6.1 CHARACTER INFORMATION

Numeric information is represented in a C++ program as integers or floating-point values. These numeric values can be single, independent values, or they can be grouped together in an array, as we saw in the last chapter. Numeric values are often used in arithmetic computations.

But in many problem solutions, we need to store and manipulate nonnumeric information, which may consist of alphabetic characters, digits, and special characters. Even though nonnumeric information can contain digits, these digits are not values generally used in arithmetic computations; instead, they are digits in an address, a phone number, or a social security number.

Recall that all information stored in a computer is represented internally as sequences of binary digits (0 or 1). In general, you do not need to be concerned about this binary representation, because the compiler and the computer perform the necessary steps to convert programs to binary and then to execute them. However, in order to work with **characters**, we need to understand more about their representation in the computer's memory.

Each character corresponds to a **binary code** value. The most commonly used binary codes are **ASCII** (American Standard Code for Information Interchange) and **EBCDIC** (Extended Binary Coded Decimal Interchange Code). The discussions that follow, assume that ASCII code is used to represent characters. Table 6-1 contains a few characters, their binary form in ASCII, and the integer values that correspond to the binary values.

Looking at the table, you can see that the character a is represented by the binary value 1100001, which is equivalent to an integer value of 97. A total of 128 characters can be represented in the ASCII code; a complete ASCII code table is given in Appendix A.

Once a character is stored in memory as a binary value, the binary value can be interpreted as a character or as an integer, as illustrated in Table 6-1. Thus, when you define variables that are to be used for storing characters, you can define them as either characters or integers. However, it is important to note that the binary representation for a character digit is not equal to the binary representation for an integer digit. From Table 6-1, we see that the binary representation of the digit 3 is equivalent to the binary representation of the integer 51. Thus, performing a computation with the character representation of a digit does not yield the same result as performing the computation with the integer representation of the digit.

Nonnumeric information can be represented by constants or by variables in programs. A character constant is enclosed in single quotes, as in `'A'`, `'b'`, and `'3'`. A variable that is going to contain a character can be defined as an integer or as a character

TABLE 6-1 Examples of ASCII Codes

CHARACTER	ASCII CODE	INTEGER EQUIVALENT
newline, \n	0001010	10
%	0100101	37
3	0110011	51
A	1000001	65
a	1100001	97
b	1100010	98
c	1100011	99

data type (`char`). Integer arrays can be used to represent a group of characters. C++ also allows the definition of a character string, but character strings are not discussed in this text.

PRACTICE!

Give integer values for the following characters using the ASCII code table from Appendix A.

1. (

2. <

3. G

4. g

Give the characters (or meaning) that correspond to each of the following binary code values.

5. 1001111

6. 0100100

7. 1110101

8. 1011110

6.2 CHARACTER INITIALIZATION AND COMPUTATIONS

The binary representation for a character can be interpreted as a character or as an integer. Similarly, the value of a small positive integer can be printed as a character or an integer, and the value of a character variable (defined using the `char` type) can be printed as a character or an integer. To print a value as an integer, use the `%i` or `%d` specifier. To print a value as a character, use the `%c` specifier. The following statements show how to print the same value as an integer and as a character:

```
//  Define and initialize variables.
int k=97;
char c=k;

//  Print the value 97 as an integer and character.
cout << "value of k: " << k
     << "; value of c: " << c << endl;
```

The output of these statements is

```
value of k: 97; value of c: a
```

In many problem solutions, either type of variable can be used to store and manipulate a character. However, there are situations in which a character should be stored in an integer variable, as illustrated in the following discussion on character I/O.

Character I/O

Although the `cout` and `cin` functions can be used to read characters using the `%c` specifier, C++ also contains special functions for reading and printing characters. The `get` function reads a character from the keyboard and returns the integer value of the character. The `put` function "prints" a character to the computer screen, and then also returns the same character. The prototype statements for these functions are

```
cout.put(char_name);
cin.get(char_name);
```

Both functions use a **text stream**, which is composed of a sequence of characters. For the `get` function, the text stream is the line entered through the keyboard. For the `put` function, the text stream is the line printed on the screen. In either case, the text stream can be separated into lines by newline characters. The end of a text stream is indicated with a special value, EOF. This special value, EOF, is a symbolic constant defined in `iostream.h`.

Executing the `put` function causes the character that corresponds to the integer argument to be written to the computer screen. If several `put` function references are made in a row, the characters are displayed one after another on the same line, until a newline character is printed. Thus, the following statements cause the characters `ab` to be shown on one line, followed by c on the next line:

```
cout.put('a');
cout.put('b');
cout.put('\n');
cout.put('c');
```

The same information could be displayed using the integer values that correspond to the characters (see Table 6-1):

```
cout.put(97);
cout.put(98);
cout.put(10);
cout.put(99);
```

In general, it is better to use character constants instead of their binary equivalents in order to make the program easier to read.

When a `get` function is executed, the next character in the current input text stream is obtained and returned as the function value. If there is no current text stream, a line of information is obtained from the keyboard. This line must be ended by pressing the return key, which corresponds to entering a newline character. Thus, when the first `get` function is executed in a program, a line of information is read from the keyboard, but only the first character in the line is returned by the function. When the next `get` function reference is executed, the second character from the line is returned by the function. Successive references continue to return additional characters, until the character returned is a newline character; this signifies that we have reached the end of the current line. Then, the next reference to the `get` function causes a new line of information to be read from the keyboard into the text stream. This processing of the text stream by lines continues until the EOF character is entered through the keyboard.

The value of the EOF character is system dependent. In the Borland Turbo C++ environment, the EOF character is entered by pressing the control (ctrl) key, and then pressing the z key while the control key is still pressed. This combination of characters is often written as ^z, but this is not the same as pressing the ^ key with the z key. In the Borland Turbo C++ environment, this EOF character is represented by the integer value −1. (Note that you cannot enter −1 for the EOF character because it will be interpreted as a minus sign followed by the digit 1.) To determine the sequence of characters that represents the EOF character on other systems, consult your instructor or a computing center consultant.

To illustrate the use of the get and put functions, consider the following program that reads characters from the keyboard and prints them to the screen. It also computes and prints a count of the characters read, including spaces and newline characters, but not including the EOF character. Note that the user is reminded of the sequence necessary to terminate the text stream.

```cpp
//-----------------------------------------------------------
//   Program chapter6_1
//  /
//   This program demonstrates the relationship
//   between a text stream and character I/O by
//   reading characters from the keyboard and then
//   printing them to the screen.

#include <iostream.h>
#include <stdlib.h>

int main()
{
   // Define and initialize variables.
  int count=0;
 char c;

  // Read, print, and count characters.
  cout << "Enter characters (^z to quit):" << endl;
   while (cin.get(c))
{
    cout.put(c);
    count++;
}

   // Print the number of characters printed.
   cout << count << " characters printed" << endl;

   // Exit program.
   return EXIT_SUCCESS;
}
//-----------------------------------------------------------
```

When the first get function is executed, a line of text is obtained from the keyboard. (Note that this line of text must be ended by pressing the return key, which is equivalent to a newline character.) The first reference to the get function returns the value of the first character of the line in the variable c, but successive executions of the get function cause additional characters on the line to be read. Therefore, you may enter as many characters as you want on a line, but processing of the program will not continue until you press the return (or enter) key.

An example of the information displayed on the screen during a sample execution of this program is the following:

```
Enter characters (^z to quit):
abcd
abcd
z
z
1w2e3r $
1w2e3r $
^z
16 characters printed
```

The first line of characters was abcd followed by a newline, the second line contained the character z followed by a newline, and the third line contained the characters 1w2e3r $ followed by a newline. The fourth line contained the EOF representation followed by a newline. The EOF representation caused the program to be terminated. Thus, a total of 5 + 2 + 9 + 1, or 17 characters, was read, but only 16 were printed.

An interesting variation of program chapter6_1 is shown in the next program, which prints each character twice and also counts the number of lines read.

```
//-------------------------------------------------------------
//  Program chapter6_2
//  /
//  This program reads characters from the keyboard
//  and prints them twice on the screen.  It also
//   counts and prints the number of lines read.

#include <iostream.h>
#include <stdlib.h>

int main()
{
   //  Define constants and variables.
  const char NEWLINE='\n';
   int count=0;
  char c;

   //  Read and print characters.
  cout << "Enter characters (^z to quit):" << endl;
   while (cin.get(c))
 {
     cout.put(c);
     cout.put(c);
    if (c == NEWLINE)    // Count the lines.
      count++;
 }
```

```
     //  Print the number of lines read.
   cout << count << " lines read" << endl;

     //  Exit program.
    return EXIT_SUCCESS;
}
//-----------------------------------------------------------
```

Sample output from this program, using the same input as for the previous program, is

```
Enter characters (^z to quit):
abcd
aabbccdd

z
zz

1w2e3r $
11ww22ee33rr   $$

^z
3 lines read
```

Note that the extra line following each line of output is generated by the duplication of the newline character at the end of each input line.

Arrays of Characters

Because a character can be stored as an integer, groups of characters can be stored in an integer array. Characters in an integer array are accessed using subscripts. Integer arrays containing characters can also be used as function arguments.

To illustrate, consider the following program, which counts the number of words in a data file containing text. Assume that words are not split between lines and that there is at least one blank between words. Also assume that the maximum number of characters on a line, including the newline character, is 100. Each character is read using the get function.

```
//-----------------------------------------------------------
//  Program chapter6_3
//  /
//  This program reads characters from a data file
//  and counts the number of words, line by line.

#include <iostream.h>
#include <fstream.h>
#include <stdlib.h>

int main()
{
   //  Define constant, variables, and prototypes.
  const char NEWLINE='\n';
  int k=0, count=0;
   char line[100];
   int word_ct(char x[], int npts);

   //  Open file.
   fstream text1;
    text1.open("text1.dat", ios::in);
```

```
   //  Read characters and count words.
   while (text1.get(line[k]))
{
   if (line[k] == NEWLINE)
   {
       if (k != 0)  // line is not blank
          count += word_ct(line,k);
      k = 0;
   }
       else  // get next character
       k++;
}

   //  Count words in the last line of data.
   if (k != 0)
       count += word_ct(line,k);

   //  Print number of words read.
   cout << count << " words read" << endl;

   //  Close file and exit program.
   text1.close();
   return EXIT_SUCCESS;
}
//------------------------------------------------------------
//  This function counts the number of words
//  in a character array.

int word_ct(char x[], int npts)
{
   //  Define and initialize variables.
   int count=0, k=0;

   //  While not at the end of the array,
   //  look for the first character of a word.
   while (k < npts)
   {
      while (k<=npts-1 && x[k]==' ')  // skip blanks
      k++;
      if (k < npts)  // not at end of array
        count++;
      while (k<npts && x[k]!=' ')  // skip letters
      k++;
   }

   //  Return word count.
   return count;
}
//------------------------------------------------------------
```

By using a function to count the number of words in a line, we were able to keep the main function short and readable. When the function finds the beginning of a word, it increments the word count. The end of the word is then determined by finding another space or reaching the end of the array. This program was executed with a data file containing the chapter opening discussion on the human genome; the output from the program gave the correct word count of 168.

Character Comparisons

Programs chapter6_1 and chapter6_2 compared the contents of the integer variable c to the EOF character. When the binary representations in c and EOF were equal, the program was terminated. In this comparison, it was clear that the values must be equal in order for the condition to be true.

However, suppose a and b are integer variables that contain characters, and you evaluate the following condition a < b. At first it may seem strange to ask if one character is less than another character, but if you consider the comparison in terms of the integer values represented by the characters, then this comparison makes sense. To find out if ‘?’ < ‘A’, simply refer to the ASCII code table in Appendix A. Because the numeric value of ‘?’ is 63 and the numeric value of ‘A’ is 65, the comparison is true.

The ordering of characters in a specific code, from low to high, is a **collating sequence**. If you study the ASCII collating sequence in Appendix A, some interesting characteristics can be observed. The character codes for the digits 0 through 9 are contiguous, the character codes for the uppercase letters A through Z are contiguous, and the character codes for the lowercase letters a through z are contiguous. Also, digits are less than uppercase letters, which are less than lowercase letters. The difference between an uppercase letter and its corresponding lowercase letter is 32. Finally, special characters are not contiguous; some special characters are before digits, others are after digits, and still others are between uppercase and lowercase letters.

Consider the following program which counts the number of digits in an input text stream:

```
//-----------------------------------------------------------
//    Program chapter6_4
//
//    This program counts and prints the
//    number of digits in an input text stream.

#include <iostream.h>
#include <stdlib.h>

int main()
{
   // Define and initialize variables.
   int count=0;
   char c;
   // Read characters and count digits.
   cout << "Enter characters (^z to quit):" << endl;
   while (cin.get(c))
      if ('0'<=c && c<='9')
         count++;
   // Print the number of digits read.
   cout << count << " digits read" << endl;

   // Exit program.
   return EXIT_SUCCESS;
}
//-----------------------------------------------------------
```

An example interaction with this program is the following:

```
Enter characters: (^z to quit)
514 East Sixth St.
```

```
            Hampton, NH 30255-0345
            ^z
            12 digits read
```

An alternative solution to this program is developed in the next section.

6.3 CHARACTER FUNCTIONS

The Standard C++ library contains a set of functions for use with characters. These functions fall into two categories. One set of functions is used to convert characters between uppercase and lowercase, and the other set is used to perform character comparisons. Each function requires an integer argument, and each function returns an integer value. The prototype statements for these functions are included in the header file ctype.h. The character comparison functions return a nonzero value if the comparison is true; otherwise, they return a zero.

tolower(c)	If c is an uppercase letter, this function returns the corresponding lowercase letter; otherwise, it returns c.
toupper(c)	If c is a lowercase letter, this function returns the corresponding uppercase letter; otherwise, it returns c.
isdigit(c)	This function returns a nonzero value if c is a decimal digit; otherwise, it returns a zero.
islower(c)	This function returns a nonzero value if c is a lowercase letter; otherwise, it returns a zero.
isupper(c)	This function returns a nonzero value if c is an uppercase letter; otherwise, it returns a zero.
isalpha(c)	This function returns a nonzero value if c is an uppercase letter or a lowercase letter; otherwise, it returns a zero.
isalnum(c)	This function returns a nonzero value if c is an **alphanumeric character** (an alphabetic character or a numeric digit); otherwise, it returns a zero.
iscntrl(c)	This function returns a nonzero value if c is a control character; otherwise, it returns a zero. (The **control characters** have integer codes of 0 through 21, and 127.)
isgraph(c)	This function returns a nonzero value if c is a character that can be printed, as opposed to a character that cannot be printed, such as a control character or a tab; otherwise, it returns a zero. (The **printing characters** have integer codes from 32 through 126.)
isprint(c)	This function returns a nonzero value if c is a printing character (includes a space); otherwise, it returns a zero.
ispunct(c)	This function returns a nonzero value if c is a printing character, with the exception of a space or a letter or a digit; otherwise, it returns a zero.
isspace(c)	This function returns a nonzero value if c is a space, formfeed, newline, carriage return, horizontal tab, or vertical tab (these characters are also referred to as **white space**); otherwise, the function returns a zero.
isxdigit(c)	This function returns a nonzero value if c is a hexa-decimal digit, which is a decimal digit or an alphabetic character A through F (or a through f); otherwise, it returns a zero.

These functions perform operations similar to some of the operations included in previous programs. In general, use a library function when possible instead of writing your own statements. Using library functions also reduces debugging time.

Let's rewrite program `chapter6_4` so that it uses a library character function to determine the number of digits in an input text stream. Also, note that an additional `include` statement is needed:

```
//-----------------------------------------
//    Program chapter6_5
//  /
//    This program counts and prints the
//   number of digits in an input text stream.

#include <iostream.h>
#include <stdlib.h>
#include <ctype.h>

int main()
{
   // Define and initialize variables.
  int count=0;
 char c;

   //  Read characters and count digits.
  cout << "Enter characters (^z to quit):" << endl;
   while (cin.get(c))
    if (isdigit(c))
      count++;

   //  Print the number of digits read.
  cout << count << " digits read" << endl;

   //  Exit program.
  return EXIT_SUCCESS;
}
//-----------------------------------------
```

The sample output with this program is same as with program `chapter6_4`.

C++ STATEMENT SUMMARY

Include character function header file

```
#include <ctype.h>
```

Declare and initialize character variable

```
char c = '*';
```

Read character from the keyboard

```
cin.get(c)
```

Display character on the screen

```
cout.put(c)
```

Debugging Notes

1. Remember that the integer representation for a character digit is not the same as the integer representation for the numerical digit.

2. Store the value read by the get function in an integer variable so that you can compare it to the EOF character.

3. Use the character library functions instead of writing similar ones yourself to reduce the debugging time of your program.

KEY TERMS

alphanumeric character	collating sequence	EOF character
ASCII code	control character	text stream
binary code	EBCDIC code	white space
character		

Problems

Data Filters. Programs called **data filters** are often used to read the information in a data file and then analyze the contents. In many cases, this data filter program is designed to remove any data errors that would cause problems with other programs that read the information from the data file. The following set of programs is designed to perform error checking and data analysis on information in a data file. Generate data files to test all features of the programs.

1. Write a program that reads a data file that should contain only integer values, and thus should contain only digits, plus or minus signs, and white space. The program should print any invalid characters located in the file, and, at the end, it should print a count of the invalid characters located.

2. Write a program that analyzes a data file that has been determined to contain only integer values and white space. The program should print the number of lines in the file and the number of integer values (not integer digits).

3. Write a program that reads a file that contains only integers, but some of the integers have embedded commas, as in 145,020. The program should copy the information to a new file, removing any commas from the information. Do not change the number of values per line in the file.

Bar Graphs. Characters can be used to print a bar graph that corresponds to a set of numerical values. For example, the following bar graph corresponds to the integers 5, 9, 2, 4, 10, 7:

5	* * * * *
9	* * * * * * * * *
2	* *
4	* * * *
10	* * * * * * * * * *
7	* * * * * * *

4. Write a function that receives an integer array and an integer variable that contains the number of integer values in the array. If all the values are between 0 and 50, print a bar graph similar to one shown above and return a value of 0; otherwise, do not print a bar graph, and return a value of 1. Assume that the corresponding function prototype statement is int bargraph_1(int count, int data[]);

Cryptography. The science of developing secret codes has interested many people for centuries. Some of the simplest codes involve replacing a character or a group of characters with another character or group of characters. To easily decode these messages, the decoder needs the "key" that shows the replacement characters. In recent times, computers have been used very successfully to decode many codes that initially were assumed to be unbreakable. The next set of problems considers simple codes and schemes for decoding them. Generate files to test the programs.

5. A simple code can be developed by replacing each character by another character that is a fixed number of positions away in the collating sequence. For example, if each character is

replaced by the character that is two characters to the right, then the letter 'a' is replaced by the letter 'c', the letter 'b' is replaced by the letter 'd', and so on. Write a program that reads the text in a file, and then generates a new file that contains the coded text using this scheme. Do not change the newline characters or the EOF character.

6. Write a program to decode the scheme presented in problem 5. Test the program using files generated by problem 5.

7. One step in decoding a simple code such as the one described in problem 5 involves counting the number of occurrences of each character. Then, knowing that the most common letter in English is 'e', the letter that occurs most commonly in the coded message is replaced by 'e'. Similar replacements are then made based on the number of occurrences of characters in the coded message and the known occurrences of characters in the English language. This decoding often provides enough of the correct replacements that the incorrect replacements can be determined. For this problem, write a program that reads a data file and determines the number of occurrences of each of the characters in the file. Then, print the characters and the number of times that they occurred. If a character does not occur, do not print it. (HINT: Use an array to store the occurrences of the characters, based on their ASCII codes.)

8. Another simple code encodes a message in text such that the true message is represented by the first letter of each word. There are not spaces between the decoded words, but the decoded string of characters can easily be separated into words by a person. Write a program to read a data file and determine the secret message stored by the sequence of first letters of the words.

9. Write a program that encodes the text in a data file using an integer array named key that contains 26 characters. This key is read from the keyboard. The first letter contains the character that is to replace the letter a in the data file, the second letter contains the letter that is to replace the letter b in the data file, and so on. Assume that all punctuation is to be replaced by spaces. Check to be sure that the key does not map two different characters to the same one during the encoding.

10. Write a program that decodes the file that is the output of problem 9. Assume that the same integer key is read from the keyboard by this file, and is used in the decoding steps. Note that you will not be able to restore the punctuation characters.

Appendix A
ASCII Character Codes

The following table contains the 128 ASCII characters and their equivalent integer values and binary values. The characters that correspond to the integers 1 through 31 have special significance to the computer system. For example, the character BEL is represented by the integer 7 and causes the bell to sound on the keyboard.

The order of the characters from low to high represents the collating sequence; it has several interesting characteristics. Note that the digits are less than uppercase letters, and uppercase letters are less than lowercase letters. Also, note that special characters are not grouped together—some are before digits, some are after digits, and some are between uppercase and lowercase characters.

CHARACTER	INTEGER EQUIVALENT	BINARY EQUIVALENT
NUL (Blank)	000	0000000
SOH (Start of Header)	001	0000001
STX (Start of Text)	002	0000010
ETX (End of Text)	003	0000011
EOT (End of Transmission)	004	0000100
ENQ (Enquiry)	005	0000101
ACK (Acknowledge)	006	0000110
BEL (Bell)	007	0000111
BS (Backspace)	008	0001000
HT (Horizontal Tab)	009	0001001
LF (Line Feed or Newline)	010	0001010
VT (Vertical Tabulation)	011	0001011
FF (Form Feed)	012	0001100
CR (Carriage Return)	013	0001101
SO (Shift Out)	014	0001110
SI (Shift In)	015	0001111
DLE (Data Link Escape)	016	0010000
DC1 (Device Control 1)	017	0010001
DC2 (Device Control 2)	018	0010010
DC3 (Device Control 3)	019	0010011
DC 4 (Device Control 4-Stop)	020	0010100
NAK (Negative Acknowledge)	021	0010101
SYN (Synchronization)	022	0010110
ETB (End of Text Block)	023	0010111
CAN (Cancel)	024	0011000

EM (End of Medium)	025	0011001
SUB (Substitute)	026	0011010
ESC (Escape)	027	0011011
FS (File Separator)	028	0011100
GS (Group Separator)	029	0011101
RS (Record Separator)	030	0011110
US (Unit Separator)	031	0011111
SP (Space)	032	0100000
!	033	0100001
"	034	0100010
#	035	0100011
$	036	0100100
%	037	0100101
&	038	0100110
' (Closing Single Quote)	039	0100111
(	040	0101000
)	041	0101001
°	042	0101010
+	043	0101011
, (Comma)	044	0101100
- (Hyphen)	045	0101101
. (Period)	046	0101110
/	047	0101111
0	048	0110000
1	049	0110001
2	050	0110010
3	051	0110011
4	052	0110100
5	053	0110101
6	054	0110110
7	055	0110111
8	056	0111000
9	057	0111001
:	058	0111010
;	059	0111011
<	060	0111100
=	061	0111101
>	062	0111110
?	063	0111111
@	064	1000000
A	065	1000001
B	066	1000010
C	067	1000011
D	068	1000100
E	069	1000101
F	070	1000110
G	071	1000111

H	072	1001000
I	073	1001001
J	074	1001010
K	075	1001011
L	076	1001100
M	077	1001101
N	078	1001110
O	079	1001111
P	080	1010000
Q	081	1010001
R	082	1010010
S	083	1010011
T	084	1010100
U	085	1010101
V	086	1010110
W	087	1010111
X	088	1011000
Y	089	1011001
Z	090	1011010
[	091	1011011
\	092	1011100
]	093	1011101
^ (Circumflex)	094	1011110
_ (Underscore)	095	1011111
' (Opening Single Quote)	096	1100000
a	097	1100001
b	098	1100010
c	099	1100011
d	100	1100100
e	101	1100101
f	102	1100110
g	103	1100111
h	104	1101000
i	105	1101001
j	106	1101010
k	107	1101011
l	108	1101100
m	109	1101101
n	110	1101110
o	111	1101111
p	112	1110000
q	113	1110001
r	114	1110010
s	115	1110011
t	116	1110100
u	117	1110101
v	118	1110110

w	119	1110111
x	120	1111000
y	121	1111001
z	122	1111010
{	123	1111011
|	124	1111100
}	125	1111101
~	126	1111110
DEL (Delete/Rubout)	127	1111111

OPERATOR PRECEDENCE

Precedence	Operation	Associativity	Page
1	( ) [ ]	innermost first	34, 46, 120
2	+ - ++ -- (type) !	right to left (unary)	24, 33, 37, 66
3	* / %	left to right	32
4	+ -	left to right	32
5	< <= > >=	left to right	65
6	== !=	left to right	65
7	&&	left to right	66
8	||	left to right	66
9	= += -= *= /= %=	right to left	30, 38

COMMON C++ FUNCTIONS

Elementary Math Functions

ceil(x)	exp(x)	fabs(x)	floor(x)
log(x)	log10(x)	pow(x,y)	sqrt(x)

Trigonometric Functions

acos(x)	asin(x)	atan(x)	atan2(y,x)
cos(x)	sin(x)	tan(x,y)	

Character Functions

isalnum(c)	isalpha(c)	iscntrl(c)	isdigit(c)
isgraph(c)	islower(c)	isprint(c)	ispunct(c)
isspace(c)	isupper(c)	isxdigit(c)	tolower(c)
toupper(c)			

COMMON NUMERIC CONVERSION SPECIFIERS

VARIABLE TYPE	OUTPUT SPECIFIER	INPUT SPECIFIER
int	%i, %d	%i, %d
float	%f, %e, %E, %g, %G	%f, %e, %E, %g, %G
double	%f, %e, %E, %g, %G	%lf, %le, %lE, %lg, %lG

C++ STATEMENT SUMMARY

Preprocessor directives

```
#include <iostream.h>
#define FILENAME "sensor1.dat"
```

Declarations

```
const double PI=3.141593
int year_1, year_2, count=0, n[]={2,4,6};
double x[25];
char c= '*';
double sinc(double x);
```

Assignment statement

```
area = 0.5*base*(height_1 + height_2);
```

I/O statement

```
cout << "The area is " << area << " square feet" << endl;
cin >> year;
sensor1 >> time >> motion;
balloon << time << altitude << velocity;
cin.get(c);
cout.put(c);
```

Program exit statements

```
return EXIT_SUCCESS;
return count;
return;
```

If statements

```
if (d <= 30)
   velocity = 4.25 + 0.00175*d*d;
else
   velocity = 0.65 + 0.12*d - 0.0025*d*d;
```

While loop

```
while (degrees <= 360)
{
   cout << degrees << "  " << degrees*PI/180 << endl;
   degrees += 10;
}
```

Do/while loop

```
do
{
   cout << degrees << "  " << degrees*PI/180 << endl;
   degrees += 10;
} while (degrees <= 360);
```

For loop

```
for (degrees=0; degrees<=360; degrees+=10)
   printf("%6.0f %9.6f \n",degrees, degrees*PI/180);
```

File open/close functions

```
fstream sensor1;
sensor1.open("sensor1.dat",ios::in);
sensor1.close();
```

Index